Plurigaussian Simulations in Geosciences

Margaret Armstrong • Alain Galli •
Hélène Beucher • Gaëlle Le Loc'h •
Didier Renard • Brigitte Doligez •
Rémi Eschard • François Geffroy

Plurigaussian Simulations in Geosciences

Springer

Margaret Armstrong
Alain Galli
Cerna
Mines-Paristech
60 boulevard Saint Michel
75272 Paris
France
margaret.armstrong@mines-paristech.fr
alain.galli@mines-paristech.fr

François Geffroy
Geovariances
49 bis avenue Franklin Roosevelt
77212 Avon
France
geffroy@geovariances.com

Hélène Beucher
Didier Renard
Gaëlle Le Loc'h
Centre de Géosciences
Mines-Paristech
35 rue St. Honoré
77305 Fontainebleau
France
helene.beucher@mines-paristech.fr
didier.renard@mines-paristech.fr
gaelle.le_loch@mines-paristech.fr

Brigitte Doligez
Rémi Eschard
IFP Energies Nouvelles
1-4 avenue de Bois Préau
92852 Rueil-Malmaison
France
brigitte.doligez@ifpen.fr
remi.eschard@ifpen.fr

ISBN 978-3-642-19606-5 e-ISBN 978-3-642-19607-2
DOI 10.1007/978-3-642-19607-2
Springer Heidelberg Dordrecht London New York

Library of Congress Control Number: 2011930790

Cover design: deblik, Berlin, Heidelberg

Printed on acid-free paper

Springer is part of Springer Science+Business Media (www.springer.com)

Preface

Although many papers have presented different aspects of truncated gaussian and plurigaussian simulations, no single work gives a comprehensive view of the theory and practical applications of this new and interesting geostatistical simulation method. Our aim in writing the book has been to encourage readers to test this method of simulating the geometry of reservoirs and ore-bodies for themselves. That is why we wrote a suite of programs called *Pluridemo* to demonstrate how the method works.

Initially this material was developed as the notes for short courses on plurigaussian simulations. The first such course was given in Fontainebleau in September 1998. Others have been given at Petrobras training centre in Rio de Janeiro in 1999 and at the IAMG annual conference in Trondheim in 1999.

The original impetus for developing the truncated gaussian method came from Georges Matheron in response to a request from the Institut Français de Pétrole, in particular from Lucien Montadert who realized the importance of being able to simulate the internal architecture of reservoirs. The plurigaussian method was a natural development from the truncated gaussian method. This cooperation between the IFP and the Centre de Géostatistique resulted in the development of the *HERESIM* program as well as practical applications for many oil companies including AGIP, Gaz de France, Petrobras and PDVSA.

Applications of the methodology to mining were initiated during a short course to engineers and geologists of Anglo American Corporation in 1999. Soon afterward a test case was carried out for Rio Tinto Ltd on data from the Rossing uranium mine.

The authors would like to thank these mining and petroleum companies for their support in developing and testing this new simulation method. Special thanks are due to Christian Ravenne formerly of the IFP for his geological insights and for his enthusiasm while developing the method.

Preface to the Second Edition

Since the book on plurigaussian simulations was first published 8 years ago, many new case-studies have been published with applications in water resources as well as petroleum and mining. Several new case-studies are included in this edition of the book. One pleasant surprise when we carried out the literature search, was to find that truncated gaussian and plurigaussian simulations are being used as the initial reservoir model for history matching (that is, for updating the reservoir model when new production data or seismic information becomes available). The reason is that two of the methods for history matching, the ensemble Kalman filter (EnKF) and the gradual deformation method, require an underlying gaussian model.

In addition to including more case-studies, the main change made since the first edition has been to add a new theory chapter (Chap. 2). When the previous edition was published our primary objective was to show reservoir engineers and geologists how the method could be used in practice. As the method is now well-established, it is important to present the theoretical aspects of the method so that mathematicians, statisticians and physicists can see its theoretical underpinnings. Those who prefer a more intuitive presentation can skip Chap. 2, and start reading at Chap. 3.

Finally the new version of the PluriDemo package designed to allow users to discover plurigaussian simulations by "playing" with them, is now available on the website. See Chap. 9 for details.

Contents

1 Introduction 1
Putting Reservoir Simulations into Context 2
Idea Behind Truncated and Plurigaussian Simulations 3
Key Steps in a Plurigaussian Simulation 6
Step 1: Choosing the Model Type 6
Step 2: Estimating the Parameter Values 6
Step 3: Generating Gaussian Values at Wells/Drill-Holes 7
Step 4: Simulating Values at Grid Nodes Given Values at Wells 7
Recent Developments 7
History Matching 8
Layout of the Book 9

2 From a Mathematical Point of View 11
Non Centered Covariance Between Two Indicators 16
Centered Covariance Between One Indicator and One Gaussian 17
Examples 17
Linear Case with One Gaussian (n = 1) 17
Non Linear Case with One Gaussian: A Gamma Process 18
Nonlinear Case with One Gaussian and a Third Order Polynomial 20
Linear Case with Two Gaussians (n = 2) 21
Linear Case with Two Gaussians and Two Categorical Constraints 23
Non Linear Case with Two Gaussians 24
Linear Case with One Gaussian Z(x) and Seismic Constraints S(x) 24
Simulating a Truncated Gaussian 26
Convergence of the Gibbs Sampler 28
The Untruncated Case 28
Convergence in the Truncated Gaussian Case 30
Markov Chains 30
Two Integral Operators 31
Stationary Distribution 32
Irreducible Chain 32

Distance Metric and Convergence 33
The Gibbs Sampler 35
Truncated Gaussian Case 36
Summary 38
Annex 1: Conditional Distribution of a Gaussian 39
Basic Properties of a Gaussian Vector 39

3 Basic Properties of Indicators 41
Spatial Covariances, Variograms and Cross-Variograms 42
Spatial Covariances 42
Variograms and Cross-Variograms 42
Variogram Properties 43
Need for a Mathematically Consistent Method 48
Transition Probabilities 49
First Type of Transition Probability 49
Second Type of Transition Probability 51
Exercises 53

4 Proportions 59
How to Calculate Vertical Proportion Curves 59
Example 1: The Ravenscar Sequence 60
Example 2: Facies with Contrasting Anisotropies in a Gold Deposit 62
Horizontal Non-stationarity 63
Choosing the Reference Level 67
Non-stationarity 72

5 Truncation and Thresholds 73
Basic Principle in the Truncated Gaussian Method 73
Defining the Thresholds 73
Transitions Between Facies 74
Link Between Thresholds and Proportions 76
Non Stationary Facies 77
Idea Behind the Plurigaussian Method 77
Link with the Proportions 79
Parameter Simplification: Use of Thresholds 79
Choice of the Partition 80
Choice of the Correlation Matrix 81
Calculating the Thresholds 83
Generalisation to Non-stationary Case 84
When Simulations Show “Prohibited” Contacts 84
Higher Dimensional Rock-Type Rules 85

6 Variograms and Structural Analysis 87
Experimental Variograms and Cross-Variograms for Facies 87

Linking the Indicator Variograms to the Underlying Variograms 88
Variograms 88
Cross-Variograms 89
Truncated Gaussian Method 89
Cross-Variograms 90
Truncated Plurigaussian Method 90
Variograms 90
Cross-Variograms 93
Generalisation to the Non-stationary Case 93
Experimental Variograms 93
Indicator Variogram Model (Truncated Gaussian Model) 95
Variogram and Cross-Variogram Model (Truncated Plurigaussian Method, two Gaussian Functions) 96
Comparing Variogram Models for Indicators and Gaussian Functions .. 96
Sill of the Indicator Model 97
Shape of the Indicator Model 97
Practical Range of the Indicator Model 97
Anisotropies 100
Variogram Fitting 101
Stationary Case 101
Non-stationary Case 102
Transition Probabilities 104

7 Gibbs Sampler 107
Why We Need a Two Step Simulation Procedure 107
Simulating Z(x) and Z(y) When There Are No Constraints 108
Simulating Z(x) and Z(y) When Z(y) Belongs to an Interval 108
Simulating Z(x) and Z(y) When Both Belong to Intervals 111
Direct Simulation Using an Acceptance/Rejection Procedure 111
Gibbs Sampler 112
Four Sample Example 112
Alternative Updating Strategies 115
Experimentally Testing Convergence 116
Burn-in Period 116
Effect of the Range on the Burn-in Period 117
The Impact of Different Parallel Runs 119

8 Case Studies and Practical Examples 121
Choosing Which Simulation Method to Use 121
Sequence-Based Pixel Models 121
Object-Based Models 122
Nested Simulations 123
Building up the Reservoir or Orebody Model 124
Step 1: Defining the Lithotypes 124

Step 2: Dividing Reservoir or Orebody into Units 124
Step 3: Defining the Reference Level 124
Step 4: Choosing the Grid Spacing 125
Petroleum Applications of the Plurigaussian Approach 126
Constructing the Lithotype Rule with Geological Constraints 126
Simulation of Reservoir with Complex Facies Transitions 126
Simulation of the Effects of Primary Diagenesis in Complex Reservoir 129
Simulating Progradational Patterns 131
Simulation of Fractures Affecting a Specific Facies 132
Testing the Impact of Simulation Parameters 132
Testing the Impact of the Rock-Type Rule 134
Handling Heterotopic Data 134
Mining Applications of the Plurigaussian Approach 138
Roll-Front Uranium Deposit 138
First Issue 139
Second Issue 142
Simulation of a Porphyry Copper Deposit: Non-sedimentary Environment 144

9 Freeware 149
Introduction 149
Installation 149
Installation Instructions for Windows Users 149
Installation Instructions for Linux Users 150
Description of *PluriDemoSimu* 150
The Choice of the Model 150
The Rock Type Rule 151
Vertical Proportions 152
Simulation Outcome 152
Limitations of the Program 154
Description of *PluriDemoVario* 154
Choice of the Model 155
Rock Type Rule 156
Proportions of Lithotypes 156
Direction for Calculation 156
Indicator Variograms 156
Description of PluriDemoSet 158
Choice of the Underlying GRF 158
The Proportions 160
Particular Case 160

References 163

Index 173

Chapter 1
Introduction

This book focuses on two methods for simulating facies and lithotypes: truncated gaussian (Matheron et al. 1987) and plurigaussian simulations (Galli et al. 1994). The first method was developed in the late 1980s for simulating the lithotypes found in oil reservoirs because it makes much more sense to simulate the geometry and internal architecture of the reservoir first, then generate suitable values of porosity and permeability once the lithotype is known. These simulations were designed for reservoirs where the lithotypes occur in a sequential order; for example, when sandstone is followed by shaly sandstone then shale. Plurigaussian simulations are a natural extension of these. They were designed to produce a much wider range of patterns and to allow for more complicated types of contacts between facies.

Figure 1.1 shows canyons along the San Juan river in Utah where phylloidal algal mounds of Pennsylvanian age are beautifully exposed. This is typical of the sedimentary oil bearing formations that can be modelled using truncated gaussian and plurigaussian methods. This series shows tabular prograding sequences of shallow carbonates and algal mounds. The carbonates can be simulated using the simpler, truncated gaussian method. The upper part of the series consisting of algal mound and intermound facies with the sandstones capping them, is more complicated. The plurigaussian method was used to simulate its structure. The details are given in Chap. 8.

Over recent years the mining industry has also started using these two approaches to simulate the facies present in orebodies. For example, the recovery during froth flotation (a mineral processing technique for concentrating sulphide ore) depends on the rock type, rather than the grades. So for both mining deposits and oil reservoirs, the message is clear: first simulate the geometry of the lithotypes or facies then simulate its properties as a function of that type. Before outlining how truncated gaussian and plurigaussian simulations work, we would like to take the time to explain why geostatistical simulations of lithofacies are important in the petroleum industry, since this motivated the development of both methods.

M. Armstrong et al., *Plurigaussian Simulations in Geosciences*,
DOI 10.1007/978-3-642-19607-2_1, © Springer-Verlag Berlin Heidelberg 2011

Fig. 1.1 Outcrops of a typical carbonate reservoirs along the San Juan River in Utah

Putting Reservoir Simulations into Context

The aim of reservoir modelling is to construct a gridded model of the reservoir containing its petrophysical properties such as porosity, permeabilities and capillary pressure, in order to simulate its behaviour during production. As the wells are generally widely spaced and as seismics only provides indirect geological information with a low resolution, the distribution of the geological heterogeneity between wells is uncertain. Geostatistical simulations are a way of quantifying the uncertainty by providing reservoir engineers with representations of the spatial distribution of reservoir heterogeneity. Fluid flow simulations can then be performed on the model or on an upscaled model to optimise the field's development.

The precise objectives vary depending on the field's stage of development. At the appraisal stage when the aim is to produce a global reservoir model rather than a detailed one, simulations are used to estimate reserves and to quantify their uncertainty. They are also used to define recovery process scenarios. At this stage only a few wells are available and so the seismics play a crucial role.

When the field has started to be developed using primary recover processes, the aim is to optimise the location of wells. At this stage more wells are available and so the geological model is constructed using the detailed well descriptions together with the seismic interpretation. Detailed reservoir characterisation studies will have been performed, and the petrophysical variables will be available from core and log analysis. Pressure measurements in wells are used to estimate the reservoir connectivity and well test information provides a way of estimating the permeability around the wells. Seismic data can also provide information about the heterogeneity within layers. The degree of detail required in the simulations (i.e. the grid size) depends on the recovery process chosen.

During the final stage of the field's development, the initial reservoir model has to be updated in order to make the recovery process more efficient. A very detailed reservoir model is then needed, for example, to optimise the drilling of wells with a complex geometry. By this time, the well database is much larger; it contains both geological and petrophysical information. Because of its low resolution the seismic information is less important. As the production history of the field is generally available, the observed oil and water flow rates can be compared with those computed from the simulation and used to modify the model iteratively until a good fit is obtained. This process is called history matching.

The complete procedure for reservoir simulation including simulating the lithofacies on a very fine grid and transforming these into reservoir properties, is described by Matheron et al. (1987). The initial geological modelling is a critical step because it ensures the consistency of the simulation from a geological point of view. Another crucial step is upscaling the reservoir properties to a coarse grid to reduce the number of cells because the capacity of most fluid flow simulators is still limited.

Finally, all the information available on the field has to be integrated into the model. Different data have been studied at different scales and have different characteristics. For example, microscopic information such as pore throat types obtained from special core analysis has to be combined with information derived from seismic campaigns having a vertical resolution greater than 20 m. Hard data such as well logs based on physical measurements in wells, have to be combined with soft information resulting from geological interpretation of the data. So simulations have to balance hard and soft constraints.

Idea Behind Truncated and Plurigaussian Simulations

The basic idea is to start out by simulating one or more gaussian variables (i.e. with a N(0,1) distribution) at every point in the study area and then use the *rock type rule* to convert these values back into lithotypes. Figure 1.2 summarises the procedure when only one gaussian is used. This is called the truncated gaussian approach. The greytone image (left) represents the gaussian value at each point in space. In the image on the right, values below −0.6 have been coloured in dark grey, indicating one facies. Values above 0.5 have been shaded white and intermediate values have been coloured light grey. These interval definitions constitute the rock type rule.

Looking at Fig. 1.2b we see an important feature of the truncated gaussian approach: the grey facies can touch the other two facies, but black and white never touch. Or if they do, it is because the pixel size is too large. If four or five facies were simulated in this way, they would occur in a fixed order, which is defined by the sequence stratigraphy.

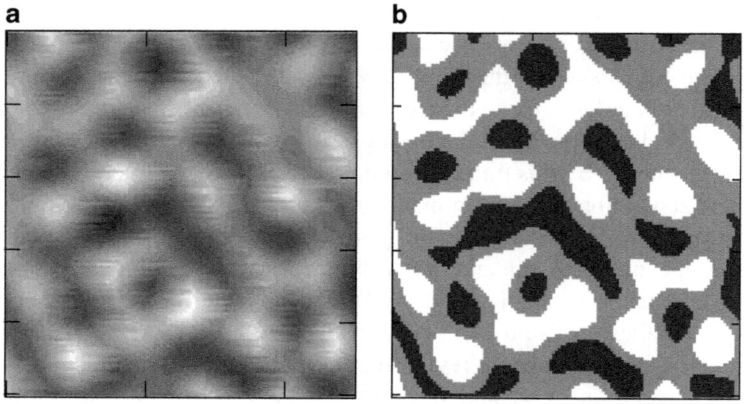

Fig. 1.2 (**a**) Simulated greytone image. Values have a N(0,1) distribution. (**b**) Same image after being truncated at the cutoffs −0.6 and 0.5. Values below −0.6 have been *shaded dark grey*, those between −0.6 and 0.5 are *coloured light grey* while values above 0.5 are shown in *white*

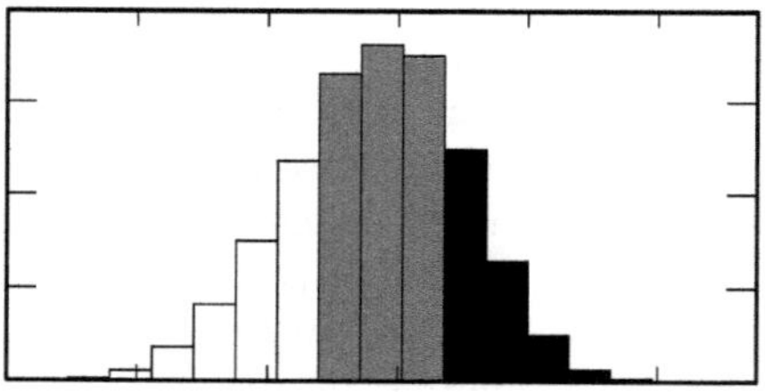

Fig. 1.3 Histogram of a standard normal distribution [i.e. N(0,1)] showing the two cut-offs, −0.6 and 0.5

Figure 1.3 shows the two thresholds, −0.6 and 0.5, on a standard normal N(0,1) distribution. It is easy to calculate the areas under the three parts of the curve (25, 45 and 30%) and hence deduce the proportion of space occupied by each facies. In fact there is a one-to-one relationship between the proportions and the thresholds, once the order has been established. In practice we compute the proportions experimentally and use these to deduce the thresholds.

In many cases, the truncated gaussian approach proves to be too restrictive; for example, if there is no natural sequence in the facies or if certain facies can be in contact with more than two facies. So it has been extended to two or more gaussians. Figure 1.4 illustrates this *plurigaussian* procedure for the case of two gaussians. At the top we see two gaussian images (simulations obtained using a gaussian random function). The one on the left has its long range in the NS direction while the other one has its long range in the EW direction. The square shown in the bottom right summarises the *rock type rule*, the rule that is used to assign each point to a facies. Values of the first gaussian, Y_1, can range from $-\infty$ to $+\infty$. They are plotted along the horizontal axis. Similarly for those of Y_2 along the vertical axis. This square is divided into three regions corresponding to different lithotypes. If $Y_2 < 0$, the rock is coded as dark grey; if $Y_2 > 0$ and $Y_1 < 0$, the rock is classified

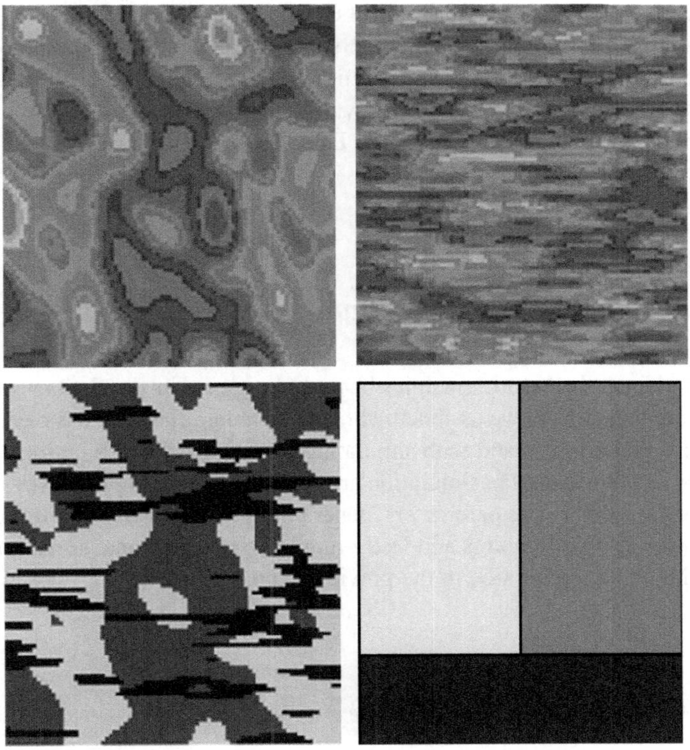

Fig. 1.4 How plurigaussian simulations work. The two images at the top are realisations of gaussian [i.e. N(0,1)] random functions with anisotropies in the NS and EW directions. The rock-type rule shown bottom *right* has been used to truncate these two images to obtain the facies or lithotypes shown bottom *left*. Note the contrast between the anisotropies in the simulated facies

as being white whereas if $Y_2 < 0$ and $Y_1 > 0$, the rock is light grey. The resulting three facies are shown in the lower left square.

The underlying idea in both truncated gaussian and plurigaussian simulations is to set up one or more simulations of standard normal random functions in the area of interest and to attribute the lithotype or facies depending on the simulated values at each point. This is done by truncating. When only one gaussian is used, the truncation is effectively defined by the values of thresholds. When two or more gaussians are used, the situation is more complex. It is represented graphically via the rock type rule.

In this book we will almost always use rectangles to divide the rock type rule into lithofacies. We do this because it makes it easier to work out the thresholds from the experimental proportions.

In most plurigaussian applications, the two gaussian random functions are independent of each other but it is also possible to use correlated gaussian RFs. When this is done the facies tend to "wrap around" each other. This makes it more difficult to compute the thresholds from the experimental proportions. It is also possible to use more than two gaussians but this makes it more difficult to compute the variograms, to estimate the parameters and to carry out the simulations.

As well as being able to use independent or correlated RFs, more advanced models can be obtained by using derivatives, or translated RFs, or any linear transform of RFs, even random ones. Even more complicated models can be obtained by combining these.

Key Steps in a Plurigaussian Simulation

Having explained the broad principles behind plurigaussian simulations we need to go into more detail. Here we assume that readers are familiar with basic geostatistics (variograms and kriging) and with simulating gaussian random functions. The four main steps in a plurigaussian simulation are choosing the appropriate type of model, estimating the values of its parameters, generating gaussian values corresponding to the lithotypes at sample points and lastly running the conditional simulation, using the gaussian values generated in the previous step.

Step 1: Choosing the Model Type

Plurigaussian simulations can be divided into several broad families, depending on the types of relations between the facies or lithotypes. For example, in some cases there is a natural sequential order among the facies. In fluvial channel reservoirs, the lithotypes – sandstone, shaly sandstone and shale – generally occur in that order because of depositional conditions. To help the reader choose the type of model that is best suited to his/her data, we provided a catalogue of examples with the freeware that is available online (see Chap. 9). When there is a clear sequential order in the lithotypes, a single gaussian usually suffices; otherwise two or more gaussians can be used. In some cases, one of the facies may appear to be a « shifted» version of another facies. Increments or derivatives of gaussians can also be used to obtain special effects.

Step 2: Estimating the Parameter Values

Two key factors control plurigaussian simulations: the thresholds at which the different gaussians are truncated and the variogram model of the underlying gaussian variable. The proportion of each facies, the "rock type" rule and the correlation between the underlying gaussian random functions determine the thresholds. Knowing the mathematical relation between the indicator variograms and the variograms of the underlying gaussian variable(s), we can find a suitable model for the underlying variogram and estimate the values of its parameters.

Step 3: Generating Gaussian Values at Wells/Drill-Holes

The facies or lithotype is known for each sample at wells (or drill-holes) but this does not tell us what the corresponding gaussian values are. We merely know that they must fall in certain domains which reduce to intervals when a rectangular partition is used. So the third step consists of generating gaussian values in the appropriate intervals and with the right properties (e.g. so as to respect the variogram model). Obviously these values are not unique. A special statistical method called a Gibbs sampler is used to generate these values.

Step 4: Simulating Values at Grid Nodes Given Values at Wells

Once the gaussian values corresponding to the facies/lithotypes have been generated at sample locations, the rest of the simulation procedure is quite straightforward. Any algorithm can be used for conditionally simulating the gaussian values at grid nodes. The last step is to convert the gaussian values at grid nodes back into facies using the rock type rule.

Recent Developments

A few theoretical advances have been made in the past 8 years since the first edition was published: Xu et al. (2006a, b) developed an approach for handling many underlying gaussians; Emery and Gonzalez (2007a, b) developed an approach for incorporating uncertainty on geological boundaries and Emery and da Silva (2009) developed a hybrid method for conditionally co-simulating continuous and categorical variables. But the main feature has been a wide range of applications in disciplines throughout the earth sciences: petroleum, mining, hydrology and environmental science. The technique has come of age in the oil industry and is now widely used for building the underlying geological model. For example, de Galard et al. (2005) used plurigaussian simulations in a study of a giant highly fractured carbonate reservoir in Iran in order to improve production; Le Maux et al. (2005) used them to study the impact of fracture modelling on reservoir performance in carbonate reservoirs; Albertao et al. (2005) used plurigaussian simulations to model an offshore reservoir in a cretaceous turbidite environment; Mubarak et al. (2009)

used them when comparing a double porosity/double permeability model to a conventional single porosity/single permeability model in fractured carbonate oil reservoirs and Pontiggia et al. (2010) used them in a study of the diagenesis of an Egyptian siliclastic reservoir. Gunning et al. (2007) used the truncated gaussian method to generate a geological model for seismic inversion.

Over the past 10 years, plurigaussian simulations have started to be used for modelling deposits in the mining industry. One of the first applications was to a granite-hosted uranium deposit (Skvortsova et al. 2000, 2002). Since then, they have been applied to a porphyry copper deposit in Chile (Carrasco et al. 2007), a roll front uranium deposit in Kazakhstan (Fontaine and Beucher 2006), a nickel laterite deposit (Rondon 2009) and to the upper portion of a diamond pipe in Botswana which is filled with sedimentary crater facies (Deraisme and Field 2006). Xu and Dowd (2008) used them when simulating rock fractures.

Applications of plurigaussian simulations are beginning in hydrogeology and environmental science. Mariethoz et al. (2006, 2009) used them to characterize aquifer heterogeneity; Cherubini et al. (2009a, b) applied them when modelling aquifer contamination; Babish (2006) included the method in a textbook designed for environmental scientists.

Probably the most significant set of new applications of plurigaussian simulations has been to construct the geological model in history matching using either an ensemble Kalman filter or gradual deformation methods, both of which require an underlying gaussian framework.

History Matching

Aanonsen et al. (2009) provide an excellent review of applications of the ensemble Kalman filter (EnKF) in reservoir engineering. The authors point out that it can be considered either as an improvement on the extended Kalman filter designed to handle more complicated data or alternatively as a sequential Bayesian inversion method. As the Kalman filter was originally developed for gaussian variables, the EnKF works best when the data are approximately gaussian.

The first step in history matching is to construct a geological model of the reservoir as input into a fluid flow simulator. As the variables in the geological model (permeability, porosity etc) depend on the lithofacies, the lithofacies are simulated first, then the porosity and permeability are simulated conditional on the lithotype. Different methods are available for simulating the lithofacies: pixel based methods such as the truncated gaussian or the plurigaussian, indicator simulations or object based methods. The advantage of using plurigaussian simulations is that they provide an underlying gaussian process on which EnKF can be applied.

Liu and Oliver (2003a, b, 2005a, b) seem to have been the first to use plurigaussian simulations to model the geology of the reservoir in history matching problems, firstly using gradient methods and then in 2005 with EnKF. Agbalaka and Oliver (2008) and Zhao et al. (2008) considered more realistic 3D reservoirs with production data. Gu and Oliver (2007) considered multiphase fluid flow; Chen and Oliver (2010) focused on improving the mass balance; Haugen et al. (2006) and Evensen et al. (2007) tested EnKF on real North Sea reservoirs.

It is important to know how this approach (plurigaussian simulations plus EnKF) compares to others and in particular whether it gives better results than the existing method, manual history matching. Aanonsen et al. (2009) commented that: Applications of EnKF to both pseudo field cases (Naevdal et al. 2005; Gu and Oliver 2005; Lorentzen et al. 2005; Gao et al. 2006) and real field cases (Skjervheim et al. 2007; Haugen et al. 2008; Evensen et al. 2007; Bianco et al. 2007) have been very encouraging. Perhaps the most important observation is that a significantly better match of production data was obtained with EnKF than using manual history matching.

Gradual deformation methods (Hu 2000) are another approach that is used in history matching. Plurigaussian simulations have been used to set up the underlying geological model. See for example, Le Ravalec-Dupin et al. (2004), Thomas et al. (2006), Gervais et al. (2009) and da Veiga and Le Ravalec (2010).

Layout of the Book

This guidebook to plurigaussian simulations is divided into nine chapters. The structure of the second edition of the book has been changed slightly compared to the first one. A new theory chapter (Chap. 2) has been included to present the method more generally from a mathematical point of view. The Gibbs sampler is used to generate gaussian values corresponding to the lithotypes/facies at the sample points. The more mathematical aspects of its convergence which were in Chap. 6 before are now at the end of Chap. 2, together with an Appendix that provides reminders on conditional gaussian distributions.

Throughout the rest of the book, a "maths-lite" approach has been used. Having said, care is required because indicator variables are deceptively simple. People have the mistaken impression that they can be treated like ordinary variables – like grades, or porosity, or the depth to a horizon. Chapter 3 reviews their properties and those of indicator variograms. Chapter 4 is on proportions. It shows how to calculate vertical proportion curves which summarise the vertical variability in the lithotypes found in oil reservoirs, and generalises this for the case where there is horizontal non-stationarity as well. Once the proportions have been modelled, the thresholds separating the facies can be calculated. Chapter 5 presents this step. The next step (Chap. 6) is to calculate the experimental indicator variograms and fit models to the variograms of the underlying gaussian variables. In Chap. 7 we illustrate how the Gibbs sampler works via several examples.

Chapter 8 focuses on applications of truncated gaussian and plurigaussian simulations, starting with several petroleum case studies. The first is on algal mounds in the Paradox Basin in Utah. The second is a synthetic case study of a reef reservoir. Two shorter studies consider prograding patterns and fracturing in one of the facies. Next we address the sensitivity of the variogram parameters and the rock-type rule on the simulations. This is followed by three new case studies: one on how to handle heterotopic data and then two mining cases. The first is on a roll-front uranium deposit in Kazakstan (i.e. a sedimentary deposit). The second is on a porphyry copper deposit in Chile. This is particularly interesting because it is not of sedimentary origin.

Chapter 9 presents three freeware programs (*PluriDemoSimu, PluriDemoVario & PluriDemoSet)* which were designed to allow readers to test plurigaussian simulations for themselves. They can be down-loaded from the web.

Chapter 2
From a Mathematical Point of View

Our objective is to simulate the facies at a set T of N target points given a data set D. The target is usually a large 2D or 3D grid, often with several hundred thousand nodes. In petroleum, the data points come from wells interpreted in lithofacies typically each foot whereas in mining studies data are core samples from drill-holes. In both cases there are usually a large number of data points too.

Let x and y denote the target points and the data locations respectively. Suppose that there are n_F facies. We model the facies via their indicators, that is, functions which take the value 1 at point x if x belongs to facies F_i and 0 otherwise. To simulate the facies, we have to sample from the conditional distribution:

$$P(1_{F_i}(x),\ x \in T | 1_{F_i}(y), i \in \{1, .., n_F\}, y \in D).$$

The two main difficulties are the availability of models for the conditional or even the unconditional distribution, and the dimension of this distribution. One classical way of solving these problems is to use indicator simulations based on kriging. But this approach has three drawbacks:

- Indicator kriging only gives a poor approximation of the conditional expectation.
- Indicator variograms and cross variograms are extremely difficult to model because of the relationships between them. This will be explained in Chap. 3.
- Often we need to address more complex problems where we have external information that is indirectly related to the facies and is defined at a different scale (for example, from seismics).

For all these reasons we chose to define the indicators by truncating a continuous multivariate random function Z(x) with n components. Let $C_i(x)$ be a partition of $\mathbb{R}^n$. A point x belongs to the ith facies F_i if and only if $Z(x) \in C_i(x)$, so

$$F_i(x) \triangleq Z(x) \in C_i(x).$$

The advantages of this model are twofold:

- Many more multivariate distributions are available for modelling continuous random function than random sets.

M. Armstrong et al., *Plurigaussian Simulations in Geosciences*,
DOI 10.1007/978-3-642-19607-2_2, © Springer-Verlag Berlin Heidelberg 2011

- The relationships between the n components $Z_j(x)$ of Z(x) are simpler than those between the facies. In particular their variograms and cross variograms are simpler to model. Furthermore the facies covariances and cross-covariances can be obtained once we know the distributions of Z.

Despite these advantages, there are still difficulties modelling and simulating these continuous random functions. Adding extra conditioning information will usually be difficult. This is why we chose to restrict the random functions that we use to be surjective functions of a of multi-gaussian random functions. The fact that we have the freedom to select the functions, gives us access to many more types of continuous random functions than the classical ones for which we know the distribution.

The problem is tractable because, as we will see below:

- With a few hypotheses on the function, we can transform the problem in terms of multi-gaussian random functions.
- Once the gaussian values at the conditioning points have been simulated, we can use the classical conditional gaussian simulation framework first developed by Matheron and Journel (see Chilès and Delfiner 1999; Lantuéjoul 2002a, b).

Another advantage of using underlying gaussians is that linear functionals such as partial derivatives or convolutions of a gaussian are still gaussian. This makes it is easier to formulate and solve the simulation problem even with additional information. Finally it is easier to find models for variograms because any negative definite function can be used as the variogram of a gaussian. Care is required when modelling the cross-variograms, but some models such as the linear coregionalisation model are known to ensure consistency. As the formulas for the variograms and cross-variograms can be derived for linear transforms of gaussians, this model is by construction consistent.

Taking advantage of these properties we now define a general setting for these simulations. As we want to be able to handle additional categorical variables and seismic attributes in addition to lithofacies, we start by defining three standardized multi-gaussian vectors, Z(x), Y(x) and S(x), which are in general correlated and which are defined below. The lithofacies are known at m_Z sample points, the categorical variables are known at m_Y points and the seismic attributes are available at a third set of m_S points. Our objective is to simulate Z(x) given these three types of data. Let

- $Z(x),\ x \in \mathbb{R}^d$ be an n-variate standardized multi-gaussian vector (i.e., the marginal and joint distributions are normal). This will be used to define the facies. The m_Z sample locations where the facies are known are denoted by x_Z. If the sample $x_\alpha \in x_Z$ belongs to the ith facies, then $Z(x_\alpha) \in \mathcal{C}_i^Z(x_\alpha)$.
- $Y(x),\ x \in \mathbb{R}^d$ be an m-variate standardized multi-gaussian vector which is correlated to Z(x). It will be used to define another categorical variable by truncation. We have information on it at m_Y points x_Y. If the sample $x_\beta \in x_Y$ belongs to the jth category, then $Y(x_\beta) \in \mathcal{C}_j^Y(x_\beta)$.Note that although we use the

same symbol for the constraints on Z and Y, the constraints are generally not in the same space unless n = m.

- $S(x),\ x \in \mathbb{R}^d$ be a p-variate standardized multi-gaussian vector correlated to Z and Y that is known at some sample locations x_S.

We assume that the three vectors have a joint multi-gaussian distribution[1]

Comments.

- We could have included the third random function S(x) in Y(x) which looks more general. However for some of the applications S(x) will be given by indirect information such as seismics, and thus it is interesting to identify it explicitly.
- To simplify the notation we will designate the constraints by the symbol $\mathcal{C}$ when there is no risk of confusion. In the cases where we want to specify one particular facies we will write $\mathcal{C}_i$ and if we need to distinguish which random function it is related to, we will add an upper index (e.g., $\mathcal{C}^Z$ to specify that it is related to Z).

Having defined the multi-gaussian random functions, we now define the functions that may be nonlinear. Let Φ, ψ and η be three functions from $\mathbb{R}^n$ onto $E \in \mathbb{R}^n$; from $\mathbb{R}^m$ onto $F \in \mathbb{R}^m$ and from $\mathbb{R}^p$ onto $H \in \mathbb{R}^p$. We assume that these functions are surjective but not necessarily bijective, that is, for any t in the arrival space we can find *at least* one point in the initial space whose image is t.

Our objective is to simulate n_F facies F_i $i = 1,.., n_F$, defined by truncating $\Phi(Z)$ at the target points subject to inequality constraints on the three variables at different points. We are now going to express this problem in three different ways.

Problem I. To simulate facies F_i defined by $\phi(Z(x)) \in \mathcal{C}^{\phi}(x)$. subject to the following constraints:

$$\begin{aligned} \phi(Z(x_Z)) &\in \mathcal{C}^{\phi}(x_Z) \\ \psi(Y(x_Y)) &\in \mathcal{C}^{\Psi}(x_Y) \\ \eta(S(x_S)) &= S^{\eta}(x_S). \end{aligned} \qquad \text{(I)}$$

Problem II. The model can be made more general by rewriting the constraints using another function Θ and a multi-gaussian B:

[1]At this level of generality no hypothesis has to be made concerning the stationarity or the correlation structure of the gaussians. They can be stationary or non stationary. Later on for practical reasons – mainly the inference of the covariances from the data but also because of the strong vertical non-stationarity of facies in sedimentary deposits linked to sequence stratigraphy – we will choose to work with second order stationary gaussians and possibly variable proportions. But from a theoretical point of view there is no need to specify these properties.

$$\Theta(B) = \begin{bmatrix} \phi(Z(x_{Z,1})) \\ \ldots. \\ \phi(Z(x_{Z,m_Z})) \\ \psi(Y(x_{Y,1})) \\ \ldots. \\ \psi(Y(x_{Y,m_Y})) \\ \eta(S(x_{S,1})) \\ \ldots. \\ \eta(S(x_{S,m_S})) \end{bmatrix}$$

While this makes it easier to write the simulation problem, it is formally equivalent to problem I. So the problem is to simulate $F_i(x)$ defined by

$$F_i = \{x \in \mathbb{R}^3;\ \phi(Z(x)) \in \mathcal{C}_i^{\phi}(x)\} \text{ given } \Theta(B) \in \mathcal{C}^{\Theta}. \tag{II}$$

Note that the $\mathcal{C}^{\Theta}$ are quite complex now because they are embedded in a higher dimension space. Although this problem looks complicated at first glance it is much simpler than it appears. As we can solve for the functions ψ and η we can also solve for the function Θ. So

$$\Theta(B) \in \mathcal{C}^{\Theta} \Leftrightarrow B \in \mathcal{C} \triangleq \Theta^{-1}(\mathcal{C}),$$

where

$$\Theta^{-1}(\mathcal{C}) = \{u \in \mathbb{R}^{nm_Y+mm_Z+pm_S};\ \Theta(u) \in \mathcal{C}\}.$$

Similarly

$$\phi(Z(x)) \in \mathcal{C}_i^{\phi}(x) \Leftrightarrow Z(x) \in \mathcal{C}_i(x),$$

With $\mathcal{C}_i \triangleq \{u \in \mathbb{R}^n; \phi(u) \in \mathcal{C}_i^{\phi}(x)\}$. Finally we see that Problem II reduces to a simpler one which we call Problem IIIa.

Problem III. In this case the problem is to simulate[2]

$$Z(x)|B \in \mathcal{C}. \tag{IIIa}$$

If need be, this problem can be rewritten in terms of Z, Y and S

$$[Z(x)|Z(x_Z) \in \mathcal{C}(x_Z), Y(x_Y) \in \mathcal{C}(x_Y), S(x_S) \in \mathcal{C}(x_S)]. \tag{IIIb}$$

[2]Here $\mathcal{C}$ refers to all the constraints ie at points x_Z, x_Y and x_S.

Note that if η is not injective, the point constraints for $S(x_S)$ are now transformed into constraints that $S(x_S)$ belongs to a set.

Problems IIIa and IIIb are far easier to solve than Problem I, because we can now use all the nice properties of gaussian random functions for simulation. The variograms will also be far simpler to model than in the initial form. The only potential drawback is that the constraints might have complex shapes and might be more difficult to compute. Now we show how to solve Problem IIIb.

Let $g(z, z_Z, z_Y, z_S)$ be the density of $(Z(x), Z(x_Z), Y(x_Y), S(x_S))$. Note that each term is itself a vector. For example Z has n components and x lies on a grid. Whatever the distribution[3] of Z, Y and S, the conditioned vector that we have to simulate has the following density (up to a normation factor that we are in general unable to compute):

$$g(z, z_Z, z_Y, z_S)1_{z_Z \in \mathcal{C}_Z} 1_{z_Y \in \mathcal{C}_Y} 1_{y_S \in \mathcal{C}_S}. \tag{2.1}$$

Using Bayes law we can rewrite it as:

$$g(z|z_Z, z_Y, z_S)h(z_Z, z_Y, z_S)1_{z_Z \in \mathcal{C}_Z} 1_{z_Y \in \mathcal{C}_Y} 1_{z_S \in \mathcal{C}_S}, \tag{2.2}$$

where h is the density of z_Z, z_Y, z_S. Alternatively it can be written as

$$g(z|z_Z, z_Y, z_S)h(z_Z|z_Y, z_S)1_{z_Z \in \mathcal{C}_Z} v(z_Y|z_S)1_{z_Y \in \mathcal{C}_Y} w(z_S)1_{z_S \in \mathcal{C}_S}. \tag{2.3}$$

Equations (2.2) and (2.3) show that we can disconnect the simulation at data points from the one at grid nodes. More precisely as soon as we have a conditional simulation at the sample locations, we can simulate Z on the grid given the values simulated at constraints points. Going further, from (2.3) we can simulate S at its own data points then Y at its data locations given the $S(x_S)$ and so on. This is independent of any hypothesis (except the existence of a density for all the variables). In the gaussian case we can go even further: (2.2) and (2.3) are saying that once we have simulated the values of the various gaussians (truncated or not) at the constraint points, we can use the well known gaussian conditional simulation methods using the preceding values as if they were known conditioning values. So now the only problem remaining is to simulate

$$h(z_Z, z_Y, z_S)1_{z_Z \in \mathcal{C}_Z} 1_{z_Y \in \mathcal{C}_Y} 1_{z_S \in \mathcal{C}_S} \tag{2.4}$$

or in the second version

$$h(z_Z|z_Y, z_S)1_{z_Z \in \mathcal{C}_Z} v(z_Y|z_S)1_{z_Y \in \mathcal{C}_Y} w(z_S)1_{z_S \in \mathcal{C}_S}. \tag{2.5}$$

For sake of completeness Annex 1 at the end of this chapter gives the proof that the conditional distributions obtained from a gaussian vector given a sub-vector are

[3] We need only assume that it has a density.

also gaussian (with the simple co-kriging as its mean and the variance of residuals as its variance).

In the multi-gaussian case using this result we see that all the three densities (h, v and w) are standardized multi-gaussians.[4] The conditional densities are also gaussian so we only need covariance matrix of the initial multi-gaussian to fully characterize these distributions. It is interesting to note that they can be computed quite easily (by a simple inversion) from covariance between facies, or categorical variables, or between one of these and the seismic data. The formulas for the two common cases are given below.

Non Centered Covariance Between Two Indicators

$$\begin{aligned} \mathrm{Cov}(1_{F_i}(x), 1_{F_j}(y)) &= P(x \in F_i, y \in F_j) \\ &= E\left[1_{F_i(x)} 1_{F_j(y)}\right] \\ &= E\left[1_{Z(x)\in \mathcal{C}_i(x)} 1_{Z(y)\in \mathcal{C}_j(y)}\right] \\ &= \int\limits_{\mathcal{C}_i(x)} \int\limits_{\mathcal{C}_j(y)} g_{z(x),z(y)}(u,v)\, dudv, \end{aligned} \tag{2.6}$$

where $g_{z(x),z(y)}(u, v)$ is the 2n-variate gaussian density of the vector (Z(x), Z(y)), so each integral is over $\mathbb{R}^n$.

For the sake of completeness we write the other non-centred covariances which are very similar

$$P(x \in {F^Y}_i, y \in {F^Y}_j) = \int\limits_{\mathcal{C}_i(x)} \int\limits_{\mathcal{C}_j(y)} g_{Y(x),Y(y)}(u, v) dudv, \tag{2.7}$$

where F^Y is the categorical variable defined by ${F^Y}_i = \{x \in \mathbb{R}^d; Y(x) \in \mathcal{C}_i(x)\}$. Note that each integral is now over $\mathbb{R}^m$.

$$P(x \in F_i, y \in {F^Y}_j) = \int\limits_{{\mathcal{C}^Z}_i(x)} \int\limits_{{\mathcal{C}^Y}_j(y)} g_{Z(x),Y(y)}(u, v) dudv. \tag{2.8}$$

Note that $g_{Z(x),Y(y)}(u, v)$ is the density of the n + m vector (Z(x), Y(y)) so the dimension of the first integral is n and the second is m.

[4]But the conditional versions are no longer standardized. See Annex 1.

Centered Covariance Between One Indicator and One Gaussian

$$\begin{aligned}\mathrm{Cov}(F_i(x), G(y)) &= E\left[1_{F_i(x)} G(y)\right] \\ &= E\left[1_{Z(x)\in\mathcal{C}_i(x)} G(y)\right] \\ &= \int\limits_{\mathcal{C}_i(x)} du \int\limits_{\mathbb{R}} s\, g_{Z(x),G(y)}(u, s) ds\end{aligned} \quad (2.9)$$

The dimension of the first integral is 1 and n for the second one.

Before studying the conditioning process we present some examples to illustrate the flexibility of the method and the variety of shapes it can produce. Geological realism will be postponed until the next chapter and the case studies.

Examples

In this section we present several examples of cases where the functions ϕ, ψ and η are linear, then several non-linear cases including one where ϕ is quadratic, which gives rise to a gamma distribution. For the linear cases we start out with the simplest case where $n = 1$ (which corresponds to the well-known truncated gaussian case). Then we consider several cases with $n = 2$, with different types of sample data.

Linear Case with One Gaussian (n = 1)

In this case there is no Y or S and $n = 1$. This is usually referred to as the truncated gaussian method. The problem is to simulate $Z(x)|Z(x_Z) \in \mathcal{C}(x_Z)$. Each simulation of Z conditioned by $Z(x_z)$ has only to be truncated in order to define the facies F_i according to the definition:

$$F_i = \left\{x \in \mathbb{R}^2;\ Z(x) \in \mathcal{C}_i(x)\right\}.$$

In this case $\mathcal{C}_i(x)$ is generally an interval, and as was mentioned earlier the $\mathcal{C}_i(x)$ for $i = 1,\ldots, n_F$ are a partition of $\mathbb{R}$ for every x. As this interval depends on x, even if we choose Z(x) to be stationary, the spatial distribution of the facies can be non-stationary. The non centered cross-covariance of facies is easy to relate to the covariance of the gaussian and to the constraints. In this case (2.9) simplifies to two integrals over $\mathbb{R}$.

$$\begin{aligned}E\left[1_{F_i}(x) 1_{F_j}(y)\right] &= E\left[1_{Z(x)\in\mathcal{C}_i(x)} 1_{Z(y)\in\mathcal{C}_j(y)}\right] \\ &= \int\limits_{\mathcal{C}_i(x)} dv \int\limits_{\mathcal{C}_j(y)} g_{Z(x),Z(y)}(u,v) du,\end{aligned} \quad (2.10)$$

where

$$g_{Z(x),Z(y)}(u,v) = \frac{1}{(2\pi)^{n/2}\det C_{x,y}} \exp\left(-\frac{1}{2}(u^t, v^t)C_{x,y}^{-1}\begin{pmatrix} u \\ v \end{pmatrix}\right)$$

and $C_{x,y}$ is the covariance Z(x) and Z(x). See Chap. 5 for more information, and in particular how to determine the covariances and the thresholds (domains $\mathcal{C}_i$), from facies data points. Equation (2.2) becomes

$$g(z|z_Z)h(z_Z)1_{z_Z \in \mathcal{C}(x_Z)}. \tag{2.11}$$

Here $g(z|z_Z)$ is the multi-variate gaussian density of dimension N (the number of target points). The mean at each target point is the simple kriging of the Z(x) using the $Z(x_Z)$, and its covariance is that of the residuals. The second and third terms give the facies information at data points, h is also a multi-variate gaussian density with mean 0 and variance 1. These two terms correspond to the density of a truncated gaussian vector to be simulated. So in order to simulate Z(x) we first simulate the $Z(x_Z)$, that is, the truncated gaussian values at data points. Then for each simulation we simulate the Z(x) at target points according to (2.11), using the preceding simulated values as conditioning points. That is, we are left with the simulation of a gaussian random function whose conditional distribution is $g(z|z_Z)$. The three examples shown in Fig. 2.1 were obtained by truncating the same gaussian random function – one with an anisotropic gaussian variogram with a parameter 100 in the horizontal direction and 50 in the vertical direction. Only the threshold intervals differ. The grey facies corresponds to the interval $[1, \infty[$ (Fig. 2.1a) while it corresponds to the interval $[0.58, \sqrt{6}[$ (Fig. 2.1b). Both have approximately 30% of grey facies. The interval $[-0.1, 0.1[$ for the grey facies is symmetric around zero (Fig. 2.1c). It has been included to help in understanding the case with two gaussians and a nonlinear transformation shown in Fig. 2.2.

Non Linear Case with One Gaussian: A Gamma Process

Let $\phi(t) = t^2$ so the facies are defined by:

$$F_i = \left\{x \in \mathbb{R}^3; Z^2(x) \in \mathcal{C}_i(x)\right\}.$$

For each value t in $\mathcal{C}_i(x) \in \mathbb{R}_+$ we have two roots $\pm\sqrt{t}$, so

$$F_i = \left\{x \in \mathbb{R}^2; Z(x) \in \pm\sqrt{\mathcal{C}_i(x)}\right\}.$$

For example, if $\mathcal{C}_i(x)$ is the interval $[a, b[$ we get:

$$F_i = \left\{x \in \mathbb{R}^2; Z(x) \in]-b, -a] \bigcup [a,b[\right\}.$$

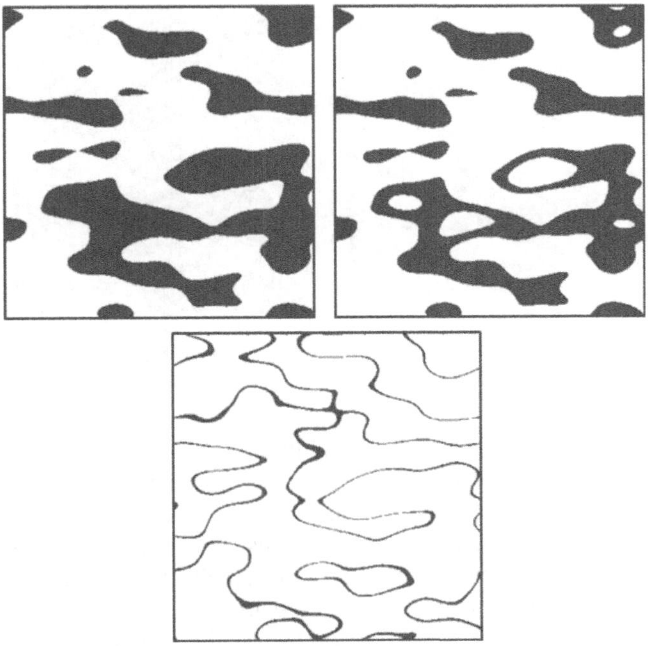

Fig. 2.1 Three examples obtained by truncating the same gaussian. The grey facies corresponds to the interval $[1, \infty[$ (top *left*), to $[0.58, \sqrt{6}[$ (top *right*) and to $[-0.1, 0.1[$ (*bottom*)

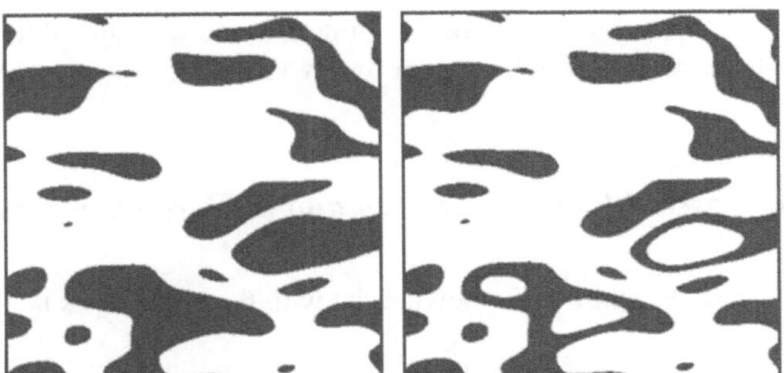

Fig. 2.2 Truncated gamma function. The underlying gaussian is the same as Figure 2.1 and the proportion of *grey* facies is very similar too. The truncation interval is $[1, \infty[$ (*left*) and $[1,6[$ (*right*)

Each facies is now defined by the union of two intervals instead of one, but except for that, we are back to the first example: linear case with one gaussian. In particular the covariances are easily expressed using (2.10) (after replacing $\mathcal{C}_i(x)$ by $-\sqrt{\mathcal{C}_i(x)} \bigcup \sqrt{\mathcal{C}_i(x)}$, the simple and cross covariances for the facies can be expressed as a sum of four covariances corresponding to the linear case with one gaussian. Two simulations of a gamma truncated process are shown in Fig. 2.2.

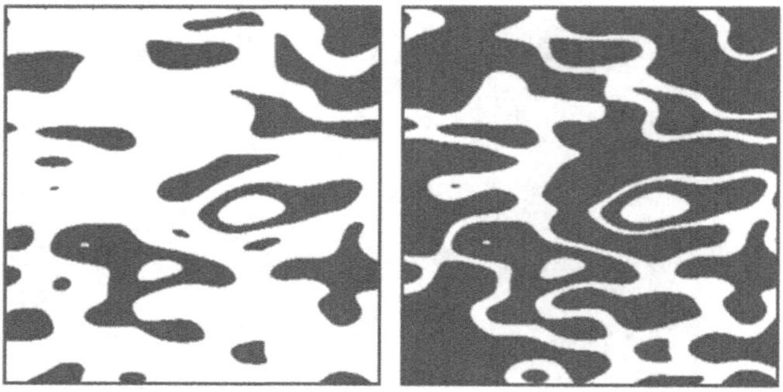

Fig. 2.3 The *grey* corresponds to $\phi(G) \leq 0$ with $t_0 = -1.08$ (*right*) and $t_0 = -0.1$ (*left*). In both cases, $t_1 = 1$ and $t_2 = 3$

The underlying gaussian is the same as in Fig. 2.1. The grey facies have been given the same proportion as in top examples. The truncation on the gamma random function is $[1,\infty[$ on the left and $[1,6[$ on the right. Comparing these to Fig. 2.1a, b the obvious difference is that the shapes are less continuous. Looking more carefully we see grey facies at the bottom left of Fig. 2.3 that was not present on Fig. 2.1. They correspond to low values of the initial gaussian.

In many cases it might be interesting to use the gaussian which defines the truncation, as an auxiliary correlated variable to simulate values in the facies (e.g., porosities or permeabilities) in order to have more continuity on the edges of the facies. This was first done by Freulon et al. (1990) for the univariate case. By doing this, we could, for example, generate a bimodal permeability distribution within one lithofacies. One population of permeabilities would correspond to high values of the underlying gaussian, and the other to low values.

Nonlinear Case with One Gaussian and a Third Order Polynomial

Consider the case of a third order polynomial with three real roots t_0, t_1, t_2:

$$\phi(t) = (t - t_0)(t - t_1)(t - t_2).$$

The truncation now depends on the roots of the polynomial. In Fig. 2.3 we use the same underlying gaussian G as in Figs. 2.1 and 2.2. The grey facies corresponds to values of G such that $\phi(G) \leq 0$, that is, to values lower than t_0 or in the interval $[t_1, t_2[$. For Fig. 2.3a we set $t_0 = -1.08$, $t_1 = 1$ and $t_2 = 3$, which gives approximately the same proportion of grey as in Figs. 2.1a, b and 2.2a. For Fig. 2.3b we changed t_0 to 0.1, while keeping the other two parameters the same. Continuing the reasoning used for the gamma case, we find that the non centered simple and cross covariances between facies are the sum of nine non centred simple and cross covariances of a classical truncated gaussian. On Fig. 2.3a the grey facies is less continuous than on Fig. 2.2a, which itself was less continuous than on Fig. 2.1b.

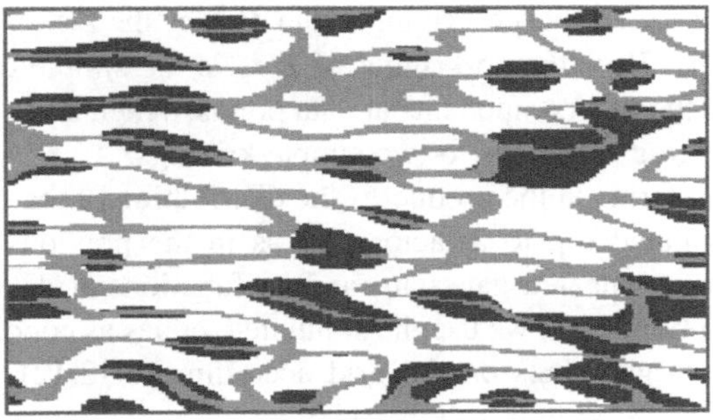

Fig. 2.4 Three facies obtained using one Gaussian and its derivative. The *grey* facies could represent meandering streams; the black one could be crevasse splays (in oil reservoirs) or washout in coal fields. Reproduced with permission

If we want to simulate petrophysical values on the grey facies using the underlying gaussian defining it as a correlated variable, we would have a different bivariate distribution than in the gamma case.

These two very simple examples show that non linear transforms of Z(x) are a convenient way to generate truncation rules that are more complex than intervals. They can provide much more freedom when fitting complex variograms of facies indicators and are useful because it is easier to understand the properties of the resulting facies by looking at the transform itself than the resulting rule on Z(x). This can also help to simulate multimodal distributions for petrophysical variables.

Linear Case with Two Gaussians (n = 2)

The only constraints are on $Z(x) = (Z_1(x), Z_2(x))$.

Classical Plurigaussian Case

The problem is to simulate $Z(x)|Z(x_z) \in \mathcal{C}(x_z)$. For each x_Z, $\mathcal{C}(x_Z)$ is a domain in $\mathbb{R}^2$ (n = 2). In most practical applications $\mathcal{C}(x_z) = I_1(x_z) \times I_2(x_z)$, where I_1 and I_2 are intervals in $\mathbb{R}$. So as soon as we have simulated Z(x) the facies at location x is given by:

$$F_i = \left\{x; Z_1(x) \in I_1^i(x) \ \& \ Z_2(x) \in I_2^i(x)\right\}.$$

Rewriting (2.2) in terms of the two components we get:

$$g(z_1, z_2|z_{1Z}, z_{2Z})h(z_{1Z}, z_{2Z})1_{z_{1Z}\in I_1(x_Z)}1_{z_{2Z}\in I_2(x_Z)} \tag{2.12}$$

where $g(z_1, z_2|z_{1Z}, z_{2Z})$ is the multi-variate gaussian distribution. The dimension of both z_1 and z_2 is that of the grid to be simulated. For every grid node x, the mean of the distributions is the simple co-kriging of the Z(x) by the $Z(x_Z)$, and its covariance is that of residuals. Here the co-kriging of Z(x) is the simple co-kriging of each component using the two components at data points. When Z_1 is independent of Z_2 this simple co-kriging simplifies to two simple krigings.

The other three terms in the product is the distribution of the truncated gaussian vector to be simulated (up to a factor). So as in the previous case we start by simulating the two truncated gaussian random functions at data points using the preceding distribution. Then we use the simulated values as conditioning points for simulating the two gaussians on the grid according to (2.12). (See Chap. 7 for details on how (2.6) is written for this case).

Up to now, we have not specified the covariance and cross covariance of Z_1 and Z_2. There is no specific constraint, provided they constitute a valid model for coregionalisation. One way to ensure this would be for example to use the linear model for coregionalisation, but there are plenty of other possibilities, e.g., by specifying a functional relationship between the two gaussians. For example Z_2 could be a convolution of Z_1 (or a convolution of a variable correlated to Z_1). Below we present a case where Z_2 is a partial derivative of Z_1

In Chap. 7 we discuss the link between the co-regionalisation structure of the Ys and that of the facies for the case of two gaussians, in more detail.

Derivative Based Plurigaussian Simulations

Here we present an example where the second gaussian is the partial derivative of the first one (Armstrong and Galli 1999). In order for the derivative of $Z_1(x)$ to exist, the covariance of $Z_1(x)$ must be twice differentiable. More general models based on other derivatives can also be used for Z_2. For example, the derivative of another random function Z_3 that is possibly correlated with Z_1 could be included:

$$Z_2(x,y) = a\frac{\partial}{\partial x}Z_1(x,y) + b\frac{\partial}{\partial y}Z_3(x,y).$$

The method is exactly the same as in the general case presented above. The only differences will be in the covariance of $Z_2(x)$ and the cross covariance with $Z_1(x)$. If we assume that the processes are stationary and if we let h be the vector of differences, then we have:

$$C_{Z_2}(h) = -a^2\frac{\partial^2}{\partial {h_1}^2}C_{Z_1}(h) - b^2\frac{\partial^2}{\partial {h_2}^2}C_{Z_3}(h) + ab\frac{\partial^2}{\partial h_1 \partial h_2}C_{Z_1,Z_3}(h)$$

$$C_{Z_1,Z_2}(h) = a\frac{\partial}{\partial h_1}C_{Z_1}(h).$$

Figure 2.4 taken from Armstrong and Galli (1999) presents an example with three facies. The grey facies represents meandering streams or rivers; the black one

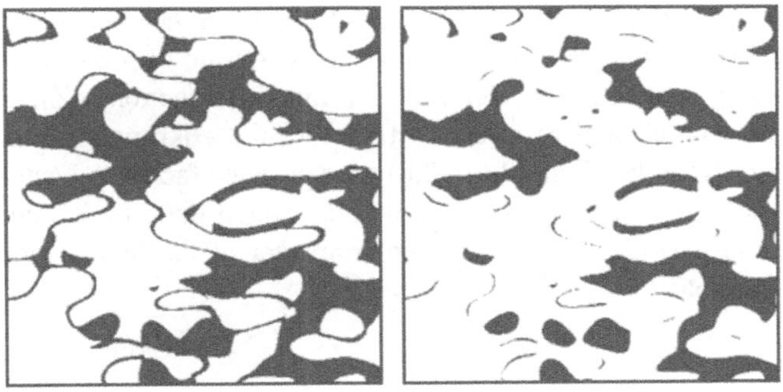

Fig. 2.5 Truncating a non-linear function of two independent gaussians, with $I_1 = [-0.2\ 2[$, $I_2 = [-0.1\ 5]$ on the *right*, and $I_1 = [-0.5\ 0.5[$, $I_2 = [0.1\ 5]$ on the *left*

could be crevasse splays in an oil reservoir or washouts in a coal measure. This sort of geometry is difficult to reproduce without using this type of model.

Linear Case with Two Gaussians and Two Categorical Constraints

There are situations where the dataset is composed of two types of categorical data at the same sample locations or at different locations. In some cases these could be related to two facies types (initial facies and diagenetic facies for instance see Pontiggia et al. 2010), whereas in others, they could be related to the same facies (for instance, different interpretations of logs). In that case we would not want to include this information in the truncation rule, because only the correlation matters.

Let the two gaussians for the facies be $Z_1(x)$ and $Z_2(x)$; let $Y_1(x)$ and $Y_2(x)$ be the two gaussians corresponding to categorical constraints. The simulation method is the same as before, except that an extra step has to be included: first simulate $Z(x_Z)$ and then simulate $Z(x)$ with the right conditional distribution which involves the $Z(x_Y)$ and the $Z(x_Z)$. For simplicity we assume that various $\mathcal{C}$ are the products of intervals. Equation (2.2) becomes:

$$\begin{aligned} & g(z_1, z_2 | z_{1Z}, z_{2Z}, z_{1Y}, z_{2Y}) \\ & \quad \times h(z_{1Z}, z_{2Z} | z_{1Y}, z_{2Y}) 1_{Z_{1z} \in I_Z^1(x_Z)} 1_{Z_{2z} \in I_Z^2(x_Z)} \\ & \quad \times v(z_{1Y}, z_{2Y}) 1_{Z_{1y} \in I_Y^1(x_Y)} 1_{Z_{2Y} \in I_Y^2(x_Y)}. \end{aligned} \qquad (2.13)$$

The conditional distribution of the $Z(x_Z)$ given the $Y(x_Y)$, is a truncated gaussian whose mean is the cokriging of each component of the Z by the known values of the Y and the covariance matrix is that of the corresponding residuals. So the simulation of Z at points x_Z is carried out using this distribution. The next step is to simulate the two components at grid nodes using their conditional distribution which is also gaussian.

From a theoretical point of view, there is little difference between this and the case without constraints on Y, but from a practical point of view we must have a consistent coregionalisation model for four random functions (Z_1, Z_2, Y_1, Y_2) which can be complicated to infer. Furthermore the cokriging starts to become quite involved. Note that from (2.6)–(2.8) we can relate facies and categorical data to the various gaussian. So provided we have enough data, the coregionalisation structure of the gaussian can be inferred from that of the facies and the second categorical variable.

Non Linear Case with Two Gaussians

Continuing the preceding discussion we show that the non linear transforms can be more complicated because both components can involve one or two gaussians. So the truncation rules will be far more complex, as in the simple example below. The conclusions found for one gaussian hold here too, except that the simple and cross variograms of facies are more complex.

Case with Two Facies

The first gaussian is the same as in the previous examples with an anisotropic gaussian variogram with $a_X = 100$, $a_Y = 50$, while the second is isotropic with a gaussian variogram with $a = 50$. To simplify further we let the two gaussians be independent

$$F_1 = \left\{ \begin{array}{l} Z_1(x) \in I^1 \\ Z_1(x) Z_2(x) \in I^2 \end{array} \right\}.$$

This case is interesting when the origin belongs to I^1. At points where Z_1 equals 0 no facies F_1 can be present unless I^2 also contains 0. So we are able to generate two types of grey facies, a massive one and a thin one. In the case on the left the thin ones are connected to the massive ones, while on the right they are isolated (Fig. 2.5).

Linear Case with One Gaussian Z(x) and Seismic Constraints S(x)

Suppose that we want to simulate three facies obtained by truncating Z(x) given seismic information S(x). For simplicity we assume that this seismic information is a vertical convolution: (in 3D x is denoted by (x_1, x_2, x_3))

$$S(x) = \int_c^d w(x_3 - y)\, G(x_1, x_2, y)\, dy. \tag{2.14}$$

Here G(x) is a standard normal distribution N(0,1) that is correlated to Z(x). The function w has been normalized so that S(x) is also N(0,1). In this very simple case with conditioning data, (2.5) simplifies (up to a constant factor) to:

$$h(z|z_S)1_{z_Z \in \mathcal{C}_Z}.$$

Here z_S stands for the values of seismics at sample locations $S(x_S)$, and h is the gaussian distribution of Z given S.

The mean of this gaussian conditional distribution is just the simple cokriging of Z knowing the seismics at data points; its covariance is the covariance of the simple kriging residual. So provided we know the covariance between Z and S we can carry out all the computation required. In order to compute this covariance, we use the classic convolution formula, but for that we need the covariance between F_i and G, as shown below.

$$\begin{aligned}C_{F_i,S}(x+h,x) &= E(1_{F_i}(x+h)S(x))\\ &= \int_c^d w(x_3+h_3-y)\mathrm{Cov}\left(1_{Z(x+h)\in\mathcal{C}_i(x+h)}, G(x_1,x_2,y)\right) dy.\end{aligned}$$

By applying (2.13) or by direct computation, we obtain:

$$\begin{aligned}\mathrm{Cov}\left(1_{Z(x+h)\in\mathcal{C}_i(x+h)}, G(x_1,x_2,y)\right) &= E\left(1_{Z(x+h)\in\mathcal{C}_i(x+h)}G(x_1,x_2,y)\right)\\ &= \int_{\mathbb{R}} sg(s)\left[N\left(\frac{b_i-\lambda(x+h,x)s}{\sigma(x+h,x)}\right)-N\left(\frac{a_i-\lambda(x+h,x)s}{\sigma(x+h,x)}\right)\right]ds,\end{aligned} \tag{2.15}$$

where

$$\begin{aligned}\lambda(x+h,x) &= C_{Z,G}(x+h,x)\\ \sigma(x+h,x) &= \sqrt{1-C_{Z,G}^2(x+h,x)}\end{aligned}$$

and $C_{Z,G}$ is the cross-covariance between Z and G and $\mathcal{C}_i(x+h) =]a_i, b_i]$.

Assuming the stationarity of the cross-covariance between Y and S we can estimate the covariance between Z and G simply by inverting equations (2.15) if we have enough information to compute an experimental cross-covariance (or cross variogram) between the facies and the seismic. Then we use the convolution formula to obtain the required covariances, C_S and $C_{Z\ S}$.

Figure 2.6 shows a synthetic example where seismics (top) were used when simulating the facies. Two different correlation coefficients were compared: a weaker one (ρ = 0.4) above and a stronger one (ρ = 0.8) below.

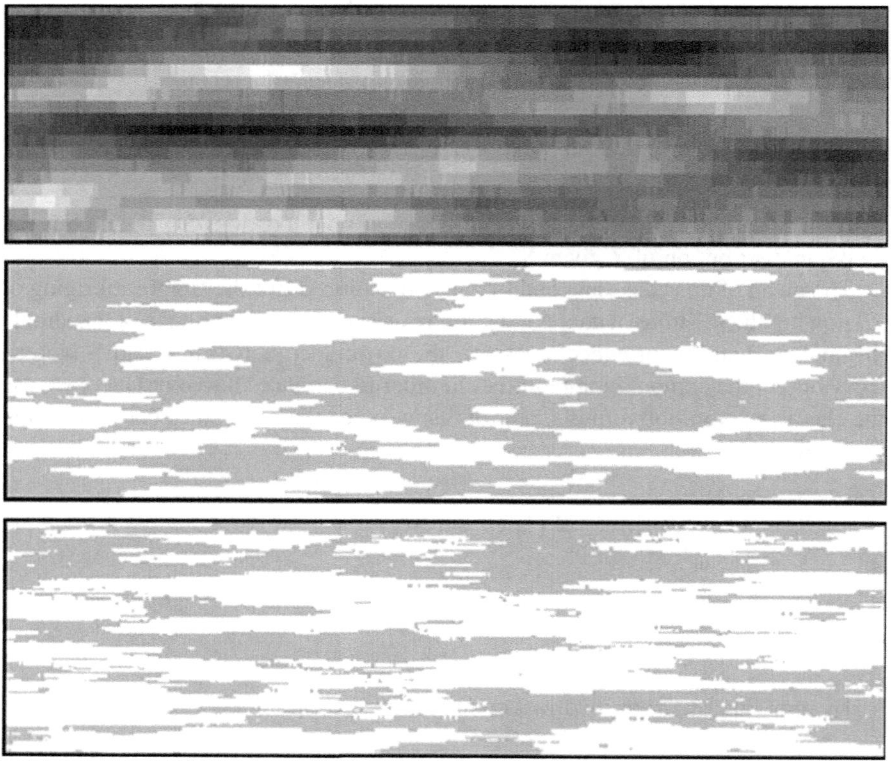

Fig. 2.6 Synthetic example where the seismics (above) were taken into account when simulating the facies. The middle panel corresponds to a correlation of 0.4, compared to 0.8 (below)

We have seen that the gaussian Z could be used as a driving factor to simulate continuous variables such as grades, porosities or permeabilities in the facies. In some cases it might be interesting to simulate the additional categorical variable Y(x). In other cases, a shift could be introduced into the correlation between the gaussian and the facies. For example, Langlais et al. (2008) used a deterministic shift in a case study on a roll-front uranium deposit, while Emery (2007a, b) used a stochastic one.

Simulating a Truncated Gaussian

There are many direct methods available for simulating gaussian random functions. The simplest are by matrix factorization, then by simulating the residuals, or by the sequential gaussian simulation where each point is simulated in turn. The problem is more difficult for truncated gaussian functions. To the best of our knowledge, no direct method exists for truncated gaussians random functions so we have to resort to MCMC.

To start we note that if $Z = (Z_1, Z_2)$ is a multivariate gaussian vector with n components that has been partitioned into n_1, n_2 then the conditional distribution of the vector Z_1 given Z_2 is gaussian. Its mean is the simple co-kriging of Z_1 by Z_2 and its covariance matrix is the covariance of residuals. So if all of the values but one are known, then we only have to simulate a truncated random variable within a known set. Many algorithms are available for doing this. The simplest based on inversion is not the most efficient. Others based on acceptation & rejection algorithms using suitable functions (Devroye 1986; Geweke 1991; Robert 1995) have a high acceptance rate. Chopin (2011) proposed an extremely fast algorithm when only one sample has to be drawn. A variant of it exists for bivariate truncated Gaussians for infinite intervals.

This suggests that the Gibbs sampler might be a good choice because it chooses one component after another and draws values according to the conditional distribution given all the others (that is, from a univariate truncated gaussian). The Gibbs sampler can be described as follows:

Given a truncated multivariate gaussian vector Z to be simulated in a set $\mathcal{C}$ we first partition the vector Z into (Z_1, Z_2) with n_1 and n_2 components respectively. Then the algorithm is:

Choose an initial vector $Z^0 = (Z_1^0, Z_2^0)$ satisfying the constraints.
Iterate

1. Draw Z_1^{n+1} following a truncated gaussian distribution in the set $\mathcal{C}$ given the values of Z_2^n
2. Draw Z_2^{n+1} following a truncated gaussian distribution in the set $\mathcal{C}$ given the values of Z_1^{n+1}

Two points in the method have to be noted:

1. The question of when to stop the iterations is treated in Chap. 7
2. The question of how to draw the two sub-vectors can be solved either by Gibbs sampling the sub-vector or in the case where n_1 (or n_2) equals 1 by drawing an univariate truncated gaussian using one of the algorithms mentioned below

The case where $n_2 = 1$ was first treated by Freulon (1994) in order to simulate conditionally truncated univariate gaussians with interval constraints, and by Le Loc'h and Galli (1997) for the bivariate truncated plurigaussian with interval constraints. But it might be worthwhile to consider different blocking cases for practical reasons, for example taking Z_1 as the vector $Z(x_Z)$ and Z_2 as $Y(x_Y)$.

Other ways to proceed could be to follow (2.5) and simulate Y first under its constraints and then Z. But as is well known (Robert and Sahu 1997; Galli and Gao 2001) the convergence rate depends strongly on the blocking and also the ordering of the components. A study of the best ordering and blocking can be made for the gaussian case and used for the plurigaussian simulation. A practical example with four conditioning samples is treated in Chap. 7.

Convergence of the Gibbs Sampler

The Untruncated Case

The Gibbs sampler is known to converge in the untruncated gaussian case. See Barone and Frigessi (1990), Amit and Grenader (1991), Roberts and Sahu (1997) or Galli and Gao (2001). In fact, the Gibbs Sampler corresponds to Gauss-Seidel iterations on the inverse of the covariance matrix. This can be proved by expressing the inverse of the covariance matrix C^{-1} as D(I−L−U) where D is the block diagonal part of C^{-1}, L and U are lower and upper block triangular matrices and I is the identity matrix. Then the iteration matrix of the Gibbs sampler can be written as $A = (I-L)^{-1}\,U$. So the rate of convergence depends on its spectral radius ρ(A). A simple proof based on relating these iterations to linear fixed point iterations is given in Galli and Gao (2001). The iteration is:

$$Y^{(n+1)} = AY^{(n)} + (I-L)^{-1}R^{(n+1)},$$

where $R^{(n+1)} = L_R V^{(n+1)}$. Here L_R is the lower triangular part of the Cholesky decomposition of D^{-1} and $V^{(n+1)}$ is a vector containing independent N(0,1) components. Consequently the covariance matrix of $R^{(n+1)}$ is D^{-1}. The equation can be written in a more compact form.

$$Y^{(n+1)} = AY^{(n)} + U^{(n+1)},$$

where the covariance matrix of $U^{(n+1)}$ is

$$\Sigma = (I-L)^{-1}D^{-1}(I-L)^{-t}.$$

Blocking Factor

Galli and Gao (2001) also demonstrated the importance of the block size used in the Gibbs sampler (i.e., in the matrices D, L and U) for geostatistical applications. The most natural choice for D is simply the diagonal terms of C^{-1}. This corresponds to blocks of size 1 × 1. Other choices are possible and these can significantly increase the speed of convergence.

To illustrate its impact, we compare the spectral radius for the 1 × 1 blocking size with that for the 2 × 2 case for the four sample example considered earlier in the chapter. In that case the covariance matrix is of the form:

$$C = \begin{bmatrix} 1 & a & a^2 & a^3 \\ a & 1 & a & a^2 \\ a^2 & a & 1 & a \\ a^3 & a^2 & a & 1 \end{bmatrix} \text{ and } C^{-1} = \frac{1}{(a^2-1)} \begin{bmatrix} 1 & a & 0 & 0 \\ a & -(a^2+1) & a & 0 \\ 0 & a & -(a^2+1) & a \\ 0 & 0 & a & 1 \end{bmatrix}.$$

In this case, it is not difficult to calculate the various matrices and hence find the spectral radii for the two block sizes and also if the points are arranged in a different order.

Block Size of 1 (Order: 1, 2, 3, 4)

$$D = \frac{1}{(a^2 - 1)} \begin{bmatrix} 1 & 0 & 0 & 0 \\ 0 & -(a^2+1) & 0 & 0 \\ 0 & 0 & -(a^2+1) & 0 \\ 0 & 0 & 0 & 1 \end{bmatrix},$$

$$A = (I - L)^{-1} U = \begin{bmatrix} 0 & b & 0 & 0 \\ 0 & b^2 & b & 0 \\ 0 & b^3 & b^2 & b \\ 0 & b^4 & b^3 & b^2 \end{bmatrix},$$

where

$$b = \frac{a}{(a^2 + 1)}$$

and hence the spectral radius ρ(A) is the largest root of the equation:

$$\lambda^2 \left[\lambda^2 - 3b^2 \lambda + b^4 \right] = 0.$$

That is, it is

$$\rho(A) = b^2 \frac{3 + \sqrt{5}}{2} < 1 \quad \text{since} \quad \max\left(b^2\right) = \max\left(\frac{a^2}{\left(a^2+1\right)^2}\right) = \frac{1}{4}.$$

Block Size of 2 (Points 1 & 2 and 3 & 4)

$$D = \frac{1}{(a^2 - 1)} \begin{bmatrix} 1 & a & 0 & 0 \\ a & -(a^2+1) & 0 & 0 \\ 0 & 0 & -(a^2+1) & a \\ 0 & 0 & a & 1 \end{bmatrix},$$

$$A = \begin{bmatrix} 0 & 0 & a^2 & 0 \\ 0 & 0 & a & 0 \\ 0 & 0 & a^2 & 0 \\ 0 & 0 & a^3 & 0 \end{bmatrix}$$

and hence the spectral radius ρ(A) is the largest root of the equation:

$$\lambda^3\left[a^2 - \lambda\right] = 0 \Rightarrow \rho(A) = \sqrt{a}.$$

Block Size of 2 (Points 1&3 and 2&4)

$$D = \frac{1}{(a^2-1)}\begin{bmatrix} 1 & 0 & 0 & 0 \\ 0 & -(a^2+1) & 0 & 0 \\ 0 & 0 & -(a^2+1) & 0 \\ 0 & 0 & 0 & 1 \end{bmatrix},$$

$$A = \begin{bmatrix} 0 & 0 & b & 0 \\ 0 & 0 & b & b \\ 0 & 0 & 2b^2 & b^2 \\ 0 & 0 & b^2 & b^2 \end{bmatrix}$$

and hence the spectral radius ρ(A) is the largest root of the equation:

$$\lambda^2\left[\lambda^2 - 3b^2\lambda + b^4\right] = 0.$$

So it is the same as in the first case.

Convergence in the Truncated Gaussian Case

In his thesis Freulon (1992) considered two iterative ways of simulating truncated gaussian random functions. The first was based on work in the field of statistical mechanics by Metropolis et al. (1953) while the second is the Gibbs sampler (see Geman and Geman 1984). Freulon (pp 58–65) proved that both procedures converge to a truncated gaussian distribution, for the total variation distance, but his demonstration of their ergodicity only works for the case where the constraint is a compact (i.e., a closed and bounded interval). Secondly it does not provide any way to compute the rate of convergence because it is the minimum of k(x, y) on D × D which is difficult to evaluate because of the denominator. In order to obtain more general results, we need to work in a more general framework, and so we introduce Markov chains.

Markov Chains

Here we give an intuitive introduction to Markov chains. For more precise and more complete descriptions see Numelin (1984), Meyn and Tweedie (1993) or Gikhman and Skorokhod (1969). A Markov chain is a chain $X^{(0)}, X^{(1)}, \ldots, X^{(n)} \ldots$ with the property that for n > 1:

$$E(X^{(n)}|X^{(n-1)}, \ldots. X^{(1)}, X^{(0)}) = E(X^{(n)}|X^{(n-1)}).$$

So a chain is fully determined as soon as we know the transition probabilities from one state to another. Here the possible states are not countable so they are characterised by an integral kernel k_n (x, y). If the kernel does not depend on the position in the chain, ie $k_n(x, y) = k(x, y)$ the chain is said to be homogeneous. We will only consider this case.

Two Integral Operators

We can define two integral operators from this kernel. If f(x) is a density then the operator $g(y) = Kf(y)$ as defined below is a density:

$$g(y) = \int_S k(x, y)\, f(x)\, dx.$$

Furthermore, if f is the density of the $(n-1)^{st}$ iterate then g(y) is the density of the nth one. Let h(x) be the conditional expectation of $t(X^{(n)})$ given $X^{(n-1)}$. Then

$$h(x) = E(t(X^{(n)})|X^{(n-1)} = x) = \int_S k(x, y)t(y)dy.$$

The two operators can be expressed in a more symmetrical way by letting

$$\tilde{k}(y, x) = k(x, y).$$

In that case

$$K_F t(y) = \int_S \tilde{k}(x, y)t(x)dx.$$

Let $K^{(n)}$ be the iterated operator, that is,

$$K^{(n)}(f) = K\left(K^{(n-1)}(f)\right).$$

Similarly let $K_F{}^{(n)}$ be

$$K_F^{(n)}(t) = K_F\left(K_F^{(n-1)}(t)\right).$$

Finally let $k^{(n)}(x, y)$ be the corresponding kernel. The operator $K^{(n)}$ expresses the relationship between the density of $X^{(p)}$ and that of $X^{(p+n)}$. In contrast $K_F{}^{(n)}$ gives the conditional expectation of a function of $X^{(p+n)}$ given $X^{(p)}$

$$E(t(X^{(n+p)})|X^{(p)}).$$

Note that for a homogeneous chain the joint distribution of the iterates $X^{(n+p)}$ and $X^{(n)}$ is:

$$g^{(n+p)}(x, y) = k^{(n)}(x, y)f^{(p)}(x) = g_y^{(n)}(x)f^{(n+p)}(y).$$

Here $f^{(p)}$ and $f^{(n+p)}$ stand for the marginal densities obtained by running the chain. The last equality holds because of Bayes Theorem.

If we let $k^*(y, x)$ be the transition kernel corresponding to reverse passage from y to x, then $g_y^{(n)}(x)$ corresponds to $k^{*(n)}(y, x)$. So we get:

$$g^{(n+p)}(x, y) = k^{(n)}(x, y)f^{(p)}(x) = k^{*(n)}(y, x)f^{(n+p)}(y). \tag{2.16}$$

Stationary Distribution

We say that a chain has a stationary distribution if the influence of the initial point $X^{(0)}$ disappears after some time. More formally a chain has a stationary distribution if $P(X^{(n)} \in A|X^{(0)})$ tends toward $\pi(A)$ for any measurable set A, whatever the initial state $X^{(0)}$. In our case this is equivalent to $k^{(n)}(x, y) \to \pi(y)$. The chain is said to have an invariant distribution π if there exists a distribution $\pi(x)$ such that

$$\pi(y) = \int_S k(x, y)\, \pi(x)dx.$$

Clearly a stationary distribution must be an invariant distribution. Assuming that $\pi(x)$ is an invariant distribution, and applying (2.16) with $p = 1$, $n = 1$, and $f(x) = \pi(x)$ we get:

$$g^{(2)}(x, y) = k(x, y)\, \pi(x) = k^*(y, x)\, \pi(y). \tag{2.17}$$

Irreducible Chain

A Markov chain is said to be ψ irreducible if for each x in the support of a measure ψ, and for each measurable set A such that $\psi(A) > 0$, there exists an integer $p \geq 1$ such that

$$P(X^{(p)} \in A | X^{(0)} = x) > 0.$$

A chain is said to be periodic if there exists an integer $d \geq 2$ and a sequence of nonempty disjoint sets $D_1, \ldots D_{d-1}$ $i = 1, \ldots (d-1)$ such that

$$\forall\, i = 1, \ldots, r-1 \quad \text{and} \quad \forall\, x \in D_i, \;\; K(x, D_j) = 1,$$

where

$$j = (1 + 1) \bmod d.$$

A chain is aperiodic if it is not periodic. Note that if a chain is irreducible with $p = 1$, then it is aperiodic.

Distance Metric and Convergence

To define convergence we first have to define a distance criterion. One of the most commonly used distances is total variation (TV). The fact that a measure μ_n converges toward a measure μ for this distance, implies that for any bounded measurable function f we have convergence of the expectation. That is,

$$\|\mu_n - \mu\|_{TV} \to 0 \quad \text{implies that} \int (\mu_n - \mu)(dx) f(x) \to 0$$

for any measurable bounded function f. This can be written as

$$E_n(f) \to E(f).$$

We have the following convergence result for Markov chains.

Theorem 1 (Tierney 1994, 1996). *If a chain is an irreductible aperiodic Markov chain with transition kernel k and an invariant distribution π then:*

$$\left\| k_F^{(n)}(x, .) - \pi(.) \right\|_{TV} \to 0.$$

Although this result is interesting, its usefulness is limited because of its generality. Secondly it does not give the rate of convergence. Stronger results can obtained in L^2 type spaces, provided one makes some additional assumptions about the compactness of the integral operators defined previously. Here we will restrict ourselves to the case where the state space S is compact. For the truncated gaussian case this excludes unbounded intervals. From a practical point of view, it means adding two additional facies with infinitesimal proportions which are not seen in the data in order to bound the interval by a large value.

For that case the previous result provides a method to estimate the speed of convergence in practice, at least for reasonably small spaces. In the case of truncated gaussian this means a reasonable number of constraints. The idea is to estimate the transition operator while running the chain as was proposed by Lantuéjoul (2002a, b). But here the chain has a continuous state space, so it is not possible to consider all possible transitions.

Given a finite partition of S, that is, a family of disjoint sets Δ_i such that:

$$S = \bigcup_{i=1}^{n} \Delta_i,$$

we consider the transition matrix:

$$P_{i,j}{}^{\Delta} = \frac{1}{|\Delta_i|} \int_{\Delta_i} \int_{\Delta_j} k(x,y)\, dx\, dy.$$

This can be estimated reasonably by running the chain for a long enough time. Then we can compute the eigenvalues of $P_{i,j}{}^{\Delta}$. So the question is now: How close are these eigenvalues to those of K.? Our assumptions are

- *The state space S is compact.*
- *The kernel k is continuous*
- *The assumptions of Theorem 1 hold.*

The first and second assumptions ensure that K will map C(S) to C(S). This is easy to verify because for continuous functions on a compact, there is no problem inverting the limit and the integral. Then using a result in Yosida (1978, p 277) we can show that this new operator K considered from C(S) to C(S) is compact. The convergence result of Theorem 1 ensures that the eigenvalues of K are lower than or equal to 1, and that it has one and only one eigenvalue 1, which is associated to the eigenvector π. This is easy to show because the set of bounded measurable functions on S includes continuous functions. Consequently

$$\operatorname*{Sup}_{f \in C[S]; |f| \leq 1} \left| \int_S \left(k^{(n)}(x,y) - \pi(y) \right) f(y)\, dy \right| \leq \left\| k^{(n)} - \pi \right\|_{TV}.$$

If there were a distribution $g \neq \pi$ such that $K_F(g) = cg$ with $c \geq 1$, then $K^{(n)}{}_F(g)$ would equal c^n g and the integral

$$\int_S (c^n g(y) - \pi(y)) dy$$

would not converge to 0. This, in turn, implies that $\|k^{(n)} - \pi\|_{TV}$ would not converge. Finally if we define

$$P_\Delta(f)(t) = \frac{1}{|\Delta_i|} \int_{\Delta_i} f(u) du \quad \text{for } t \text{ in } \Delta_i.$$

We see that when

$$|\Delta| \to 0, P_\Delta f(t) \to f(t).$$

Finally it can be shown that this corresponds to a Galerkin approximation. See Chatelin (1983, pp 170–174). That guarantees the convergence of P^Δ toward k as $|\Delta|$ tends toward 0. Interestingly, the rate of convergence depends on the derivability of k (see exercise 4.39 p 191 in Chatelin (1983). Furthermore the convergence of eigenvalues is also ensured and bounds are known.

The Gibbs Sampler

Let $\pi(x)$ be the density of the vector $(X_1,\ldots X_n)$, and let $\pi(x_i|x_{i-},x_{i+})$ be the conditional density of X_i given (X_{i-}, X_{i+}) where

$$X_{i-} = (X_1, \ldots, X_{i-1}) \quad \text{and} \quad X_{i+} = (X_{i+1}, \ldots X_n).$$

By convention

$$X_{0-} = \emptyset \quad \text{and} \quad X_{n+} = \emptyset.$$

The Gibbs sampler modifies all the coordinates of the vector, one after another, using the newly simulated one and the former ones. Note that X_i may be either a single component or a block of components. In the latter case we should speak of the block Gibbs sampler, but for simplicity we will call it a Gibbs sampler even if we are using a blocking factor.

The transition kernel for one iteration of the Gibbs sampler from x to y is:

$$k(x,y) = \prod_{i=1}^{n} \pi(y_i|y_{i-}, x_{i+}).$$

The positivity condition is said to hold if the support of π is the product of the supports of its marginal distributions. This turns out to be true in our case provided the constraints are products of unions of intervals.

Using this positivity condition we can establish the Hammersley Clifford decomposition (Besag 1974). It is easy to show that

$$\begin{aligned}\pi(x) &= \pi(x_1|x_{1-}, x_{1+})\ \pi(x_{1-}, x_{1+}) \\ &= \frac{\pi(x_1|x_{1-}, x_{1+})}{\pi(y_1|x_{1-}, x_{1+})}\ \pi(y_1, x_{1-}, x_{1+}).\end{aligned}$$

By repeating this operation on $\pi(y_1, x_{1-}, x_{1+})$ we obtain:

$$\pi(x) = \frac{\prod_{i=1}^{n} \pi(x_i|y_{i-}, x_{i+})}{\prod_{i=1}^{n} \pi(y_i|y_{i-}, x_{i+})} \pi(y) \quad \text{(Hammersley Clifford)}$$

The denominator is k(x,y) and the numerator is the kernel k*(y, x) corresponding to one iteration of the Gibbs sampler from y to x in the reverse order. So we can rewrite the Hammersley Clifford decomposition in the following way:

$$\pi(x)k(x, y) = \pi(y)k^*(y, x). \tag{2.18}$$

Integrating this gives

$$\begin{aligned} \int \pi(x)k(x, y)dx &= \int \pi(y)k^*(y, x)dx \\ &= \pi(y) \int k^*(y, x)dx \\ &= \pi(y). \end{aligned}$$

Consequently π is an invariant distribution of the Markov chain with transition kernel k(x,y) and also of the Markov chain with transition kernel k*(y, x). Equation (2.18) could be obtained directly from (2.16) without the positivity assumption but assuming that π is the stationary distribution of the chain.

Truncated Gaussian Case

In the truncated gaussian case

$$\pi(x) = \kappa\, g(x) 1_D(x),$$

where κ is the normalisation constant and $1_D(x)$ is the indicator of constraints. We will only consider constraints which are products of unions of intervals. So the positivity condition holds because the conditional distributions $\pi(x_i|x_{i-}, x_{i+})$ are truncated gaussians whose domain is the union of intervals corresponding to the constraint for the ith component. Let the inverse of the covariance matrix of X be

$$C^{-1} = D(I - L - U),$$

where D is the block diagonal part of the inverse of C, and U and L are the upper and lower triangular parts.

Using the results for the non truncated gaussian case, it is easy to show that the transition kernel is:

$$k(x, y) = k(x) e^{-\frac{1}{2}(y-Ax)^t \Sigma (y-Ax)} 1_D(y) 1_D(x),$$

where k(x) is a normation factor whose value depends on x and

$$A = (I - L)^{-1}U \quad \text{and} \quad \Sigma = (I - L)^t D(I - L).$$

Similarly,

$$k^*(y, x) = k^*(y)e^{-\frac{1}{2}(x-\tilde{A}y)^t\tilde{\Sigma}(x-\tilde{A}y)} 1_D(y) 1_D(x),$$

where

$$\tilde{A} = (I - U)^{-1}L \text{ and } \tilde{\Sigma} = (I - U)^t D(I - U).$$

It is obvious that the conditions for Theorem 1 hold so the Gibbs sampler converges toward our candidate distribution π, for the total variation distance.

Then the kernels are continuous in the support of π, and if D is compact the two operators K and K_F corresponding to the truncated gaussian case define compacts operators from C(D) to C(D). So we can apply the approximation method described earlier to estimate the eigenvalues, and to estimate the speed of convergence experimentally. It is also interesting to note that provided the constraints are the product of compact intervals, the minorisation condition holds.

Definition. *We say that the kernel k(x,y) satisfies the minorisation condition if there exists a density ψ, and a parameter $0 < \rho < 1$ such that*

$$\forall x \in D \, \forall y \in D \; k(x,y) \geq \rho \; \psi(y).$$

Theorem 2. *If the constraints D are compact the kernel k(x,y) of the transition operator satisfies the minorisation condition.*

Proof. From above

$$k(x,y) = k(x)e^{-\frac{1}{2}(y-Ax)^t\Sigma(y-Ax)} 1_D(y) 1_D(x).$$

As the proof depends on the properties of convex sets, we start out by reviewing some basic facts about them. A point v in a convex is an extremum if it cannot be written as a convex combination of other points in the convex. In a polyhedral convex such as y−A(D), the extremal points are just the vertices of the polyhedra.

Because D is compact and k(x) is continuous, it is bounded below by a parameter c which is greater than 0 because $k(x) \approx P(D - Ax)$, and its minimum is reached at at least one point x_0 and $P(D - Ax_0) > 0$. In order to minimise the exponential term we have to maximise the exponent.

Let $u = y - Ax$. This is equivalent to maximising

$$||u||_K^2 = u^t \, \Sigma u$$

subject to the constraints

$$u \in y - A(D).$$

As the function is continuous and as the set y−A(D) is compact, the minimum is attained. Because $\Sigma \geq 0$, the function is strictly convex.

As is well known, the maximum of a convex function on a convex is reached at extremal points of the convex. See for example Rockafellar (1970). Now each extremal point of y−A(D) is of the form y−A(x_i) where x_i is an extremum in D. There are 2^n extremal points x_i in D, they are the vertices of the hyper-rectangle. Let T_i be the points y whose corresponding extremal point is x_i. By definition D is the union of the T_i and we can write:

$$k(x,y) > c \sum_i 1_{T_i}(y) e^{-\frac{1}{2}(y-Ax_i)^t \Sigma (y-Ax_i)}.$$

If we let $\tilde{f}(y)$ denote

$$\tilde{f}(y) = c \sum_i 1_{T_i}(y) e^{-\frac{1}{2}(y-Ax_i)^t \Sigma (y-Ax_i)},$$

we find that

$$1 = \int_D k(x,y)dy > \int_D \tilde{f}(y)dy > 0.$$

If we let ρ be the integral of $\tilde{f}$ over D we can define the density $f(y) = \tilde{f}(y)/\rho$. We find that:

$$\forall y \in D, \forall x \in D \; k(x,y) > \rho f(y) \quad \text{with} \quad \rho < 1.$$

This minorisation condition ensures the convergence of the Gibbs sampler in L^1. See for example, Roberts and Polson (1994).

Summary

The Gibbs sampler is a very powerful simulation method. It is not difficult to establish its convergence in the truncated gaussian case. However it is far more difficult to find its rate of convergence experimentally, even in the case of compact constraints which considerably simplify the problem. For this case we proposed a finite rank operator which approximates the transition operator (which is compact).

Annex 1: Conditional Distribution of a Gaussian

Basic Properties of a Gaussian Vector

Let Z be an arbitrary gaussian vector of dimension p + n. Assume its covariance matrix Σ has an inverse Σ^{-1}. Its density is:

$$g(w_1, w_2, \ldots w_{p+n}) = \frac{1}{(2\pi)^{(p+n)/2}|\Sigma|} \exp\left\{-\frac{1}{2}(w-m)^t\, \Sigma^{-1}(w-m)\right\},$$

where w and m are column vectors with m = E(Z) and where $|\Sigma|$ is the determinant of the covariance matrix.

We split the vector Z into two parts X and Y, of dimension p and n respectively. These correspond to the p data points and the n target points to be simulated. Partitioning the covariance matrix in the same way we obtain:

$$\Sigma = \begin{pmatrix} \Sigma_{XX} & \Sigma_{XY} \\ \Sigma_{XY}^{\ t} & \Sigma_{YY} \end{pmatrix},$$

where Σ_{XX} and Σ_{YY} are the covariance matrices of X and Y, respectively, and Σ_{XY} is the cross covariance matrix between X and Y. Let Λ be the vector defined by

$$\Lambda = \Sigma_{XX}^{-1}\Sigma_{XY}.$$

It is easy to see that its ith column consists of the weights for the minimum variance linear estimate of the ith element of Y given the vector X as data. Let R be the covariance matrix of the residuals when estimating Y given X; that is, $R = \mathrm{cov}(Y - \Lambda^t X)$. It is well known[5] that the inverse of the covariance matrix Σ^{-1} is

$$\Sigma^{-1} = \begin{bmatrix} \Sigma_{XX}^{\ -1} + \Lambda R^{-1}\,\Lambda^t & -\Lambda R^{-1} \\ -R^{-1}\,\Lambda^t & R^{-1} \end{bmatrix}.$$

If we let u and v be vectors containing the first p and the remaining n components of w then

$$(u^t v^t)\ \Sigma^{-1}\begin{pmatrix} u \\ v \end{pmatrix} = (v - \Lambda^t u)^t R^{-1}\,(v - \Lambda^t u) + u^t\Sigma_{XX}^{-1}u.$$

Now we use this decomposition to express the gaussian density given earlier as the product:

---

[5]See for example Galli and Gao (2001).

$$g(u,v) = k\, g_u(v)\, g(u),$$

where the normation constant k is just

$$k = \frac{|\Sigma_{XX}||R|}{|\Sigma|},$$

$$g(u) = \frac{1}{(2\pi)^{p/2}|\Sigma_{XX}|} \exp\left\{-\frac{1}{2} u^t \sum\nolimits_{XX}^{-1} u\right\}$$

and

$$g_u(v) = \frac{1}{(2\pi)^{n/2}|R|} \exp\left\{-\frac{1}{2}(v - \Lambda^t u)^t R^{-1} (v - \Lambda^t u)\right\}.$$

Both are normal distributions; the first has zero mean and covariance Σ_{XX}, the second has mean $\Lambda^t u$ and covariance R. As their integrals are equal to 1, the normation factor k turns out 1, which was not obvious initially. The first one g(u) is the distribution of the vector X. The second is the conditional density of Y given X. It is still gaussian but its mean now equals the minimum variance estimate of Y given X, and its covariance matrix is R, the covariance matrix of the residuals. Some typical examples are:

$$Z = (Z(x_Z), Z(x)) \Rightarrow X = Z(x_Z), Y = Z(x)$$
$$Z = (Z(x_Z), Z(x), Y(x_Y)) \Rightarrow X = (Z(x_Z), Y(x_Y))$$
$$Z = (Z(x_Z), Z(x), Y(x_Y), S(x_S)) \Rightarrow X = (Z(x_Z), Y(x_Y), S(x_S)).$$

Except when working with univariate truncated gaussians, the matrix Σ is the simple co-kriging matrix. The Λ are the co-kriging weights, u stands for the data points so $\Lambda^t u$ is the co-kriging estimate at locations x and R is the matrix of the covariance of the co-kriging residuals. See Chap. 7 for more detail.

Chapter 3
Basic Properties of Indicators

The aim of this section is to review so me of the basic properties of indicators that will be needed later. We let F be a random set (i.e. all the points belonging to one particular facies). Let $\overline{F}$ be its complement (i.e. the points that do not belong to F). The indicator function for the set F takes the value 1 at all the points inside F; it takes the value 0 elsewhere. This indicator function is denoted by $1_F(x)$. In the Fig. 3.1, the set F has been shaded. Its complement $\overline{F}$ includes all of the non-shaded area. The set F could be any shape or form, and need not be a single piece. It could be split into several parts.

The first property of indicator functions relies on the fact that inside F its indicator takes the value 1 whereas the indicator of its complement takes the value zero. Outside F the converse is true. That is, the union of F and its complement fills the whole space. Consequently,

$$1 = 1_{F \cup \overline{F}}(x) = 1_F(x) + 1_{\overline{F}}(x)$$

One property of a geological facies that can be measured experimentally is the proportion of space that it occupies. If we let $P_F(x)$ be the probability that point x lies in F, then it is equal to the mean or mathematical expectation of $1_F(x)$. That is,

$$P_F(x) = E[1_F(x)]$$

As the random function $1_F(x)$ takes the values 0 or 1, its expected value must lie between these two values.

$$0 \leq P_F(x) \leq 1$$

Moreover, because of the linearity of taking expectations in,

$$P_F(x) + P_{\overline{F}}(x) = 1$$

Now we calculate its variance. Since $[1_F(x)]^2 = 1_F(x)$,

$$\mathrm{Var}\,[1_F(x)] = E[1_F(x)] - \{E[1_F(x)]\}^2 = P_F(x)(1 - P_F(x)) \leq 0.25$$

M. Armstrong et al., *Plurigaussian Simulations in Geosciences*,
DOI 10.1007/978-3-642-19607-2_3, © Springer-Verlag Berlin Heidelberg 2011

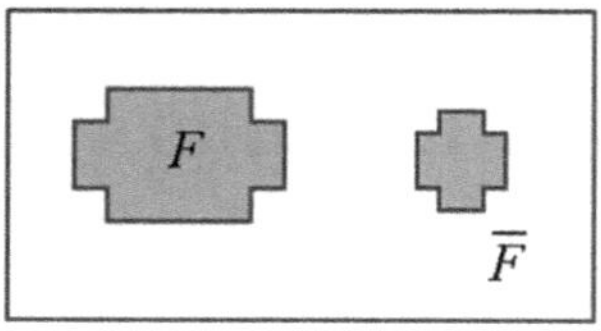

Fig. 3.1 Set F (*shaded grey*) and its complement $\overline{F}$ in *white*

Spatial Covariances, Variograms and Cross-Variograms

In this section we focus on the spatial relation between indicators by studying their centred and non-centred covariances, then their variograms and cross-variograms. After defining these concepts we derive their key properties. The first step is to study the relationship between the facies at point x and the one at y, firstly via the (non-centred) spatial covariance $C_F(x, y)$ and then via the variogram $\gamma_F(x, y)$.

Spatial Covariances

Let x and x + h be any two points. Their non-centred covariance is defined as:

$$\begin{aligned} C_F(x, x+h) &= E[1_F(x)\, 1_F(x+h)] = P[1_F(x) = 1 \,\&\, 1_F(x+h) = 1] \\ &= P[x \in F \,\&\, x+h \in F] \end{aligned}$$

It measures the probability that both points lie inside F (Fig. 3.2). Similarly the centred covariance $\sigma_F(x, x+h)$ and the centred cross covariance between facies F_i and F_j, denoted by $\sigma_{F_iF_j}(x, x+h)$ are respectively defined as

$$\begin{aligned} \sigma_F(x,\ x+h) &= E\{[1_F(x) - P_F(x)]\,[1_F(x+h) - P_F(x+h)]\} \\ &= C_F(x, x+h) - P_F(x)P_F(x+h) \\ \sigma_{Fi\,Fj}(x,\ x+h) &= E\{[1_{Fi}(x) - P_{Fi}(x)][1_{Fj}\,(x+h) - P_{Fj}(x+h)]\} \\ &= C_{Fi\,Fj}(x, x+h) - P_{Fi}(x)P_{Fj}(x+h) \end{aligned}$$

Variograms and Cross-Variograms

Let A(x) be an arbitrary variable. It could be an ordinary variable Z(x) or an indicator $1_F(x)$. Its variogram, denoted by γ_A, is defined as:

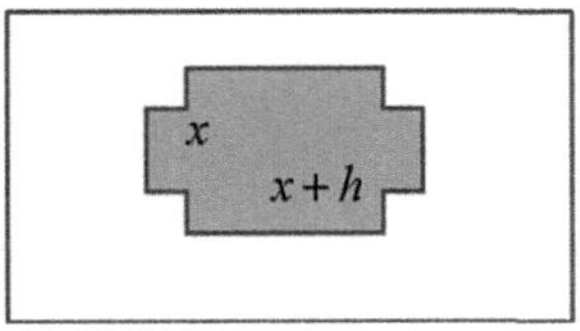

Fig. 3.2 Two points, x and x + h, lying in facies F

$$\begin{aligned}\gamma_A(x, x+h) &= \frac{1}{2}\mathrm{Var}\,[A(x) - A(x+h)]\\ &= \frac{1}{2}\left\{E([A(x) - A(x+h)]^2) - (E[A(x) - A(x+h)])^2\right\}\end{aligned}$$

Consequently the indicator variogram of the facies F, denoted by γ_F, is just:

$$\begin{aligned}\gamma_F(x, x+h) &= \frac{1}{2}\mathrm{Var}\,[1_F(x) - 1_F(x+h)]\\ &= \frac{1}{2}\left\{E([1_F(x) - 1_F(x+h)]^2) - (E[1_F(x) - 1_F(x+h)])^2\right\}\end{aligned}$$

Note that if the indicator is stationary, the second term disappears. The cross variogram between facies F_i and F_j is defined as:

$$\gamma_{F_iF_j}(x, x+h) = \frac{1}{2}E\left\{[1_{F_i}(x) - 1_{F_i}(x+h)]\left[1_{F_j}(x) - 1_{F_j}(x+h)\right]\right\}$$

When this product is expanded, two of the four terms disappear because 1_{F_i} and 1_{F_j} cannot both take the value 1 at the same point since we have one and only one facies at each point. This gives:

$$\gamma_{F_iF_j}(x, x+h) = -\frac{1}{2}\left\{E\left[1_{F_i}(x)\ 1_{F_j}(x+h)\right] + E\left[1_{F_j}(x)\ 1_{F_i}(x+h)\right]\right\}$$

Variogram Properties

Property 1

As the indicators can only take the values 0 or 1, the values of the variogram must satisfy the inequality:

$$\gamma_F(x, x+h) = \frac{1}{2}\mathrm{Var}[1_F(x) - 1_F(x+h)] \leq 0.5$$

One consequence of this is that indicator variograms must be bounded. So a power function could never be an appropriate model for an indicator variogram.

Property 2: Variograms for two facies (stationary case)

$$\gamma_F(h) = \gamma_{\overline{F}}(h)$$

Proof. By definition the variograms of F and $\overline{F}$ are

$$\gamma_F(h) = \frac{1}{2}\,E[1_F(x+h) - 1_F(x)]^2 \text{ and } \gamma_F(h) = \frac{1}{2}\,E[1_F(x+h) - 1_F(x)]$$

The first result follows because

$$1_F(x+h) - 1_F(x) = -\left(1_{\overline{F}}(x+h) - 1_{\overline{F}}(x)\right)$$

Property 3: Covariances for two facies

If F is a stationary random set, its variogram depends on the vector between the two points, but not on their positions, and similarly for the covariances. The next property follows from this.

$$\sigma_{\overline{F}}(h) = \sigma_{\overline{F}}(h) = -\,\sigma_{F\overline{F}}(h) = -\,\sigma_{\overline{F}F}(h)$$

Proof. By definition, we have

$$\sigma_F(h) = E\{[1_F(x+h) - P_F(x+h)][1_F(x) - P_F(x)]\}$$

Because of the stationarity, $P_F(x+h) = P_F(x) = P_F$ and similarly for its complement. So:

$$\begin{aligned}
\sigma_F(h) &= E[1_F(x+h) - P_F]\,[1_F(x) - P_F] \\
\sigma_{\overline{F}}(h) &= E[1_{\overline{F}}(x+h) - P_{\overline{F}}]\,[1_{\overline{F}}(x) - P_{\overline{F}}] \\
\sigma_{F\overline{F}}(h) &= E[1_F(x) - P_F]\,[1_{\overline{F}}(x+h) - P_{\overline{F}}] \\
\sigma_{\overline{F}F}(h) &= E[1_{\overline{F}}(x) - P_{\overline{F}}]\,[1_F(x+h) - P_F]
\end{aligned}$$

Moreover

$$\begin{aligned}
&1_F(x+h) - P_F = 1 - 1_{\overline{F}}(x+h) - P_F = -\left(1_{\overline{F}}(x+h) - P_{\overline{F}}\right) \\
&1_F(x) - P_F = -\left(1_{\overline{F}}(x) - P_{\overline{F}}\right)
\end{aligned}$$

Substituting these relations into the covariance formulas gives the second result.

Comments

There are three practical consequences of these results. Firstly two complementary random sets are forcibly correlated. They cannot be independent of each other.

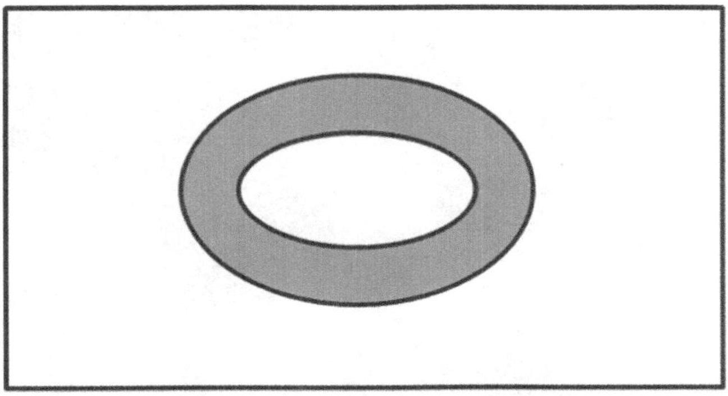

Fig. 3.3 The facies F and its complement do not have the connectivity but their indicator covariances are identical

Secondly their cross covariance is the same as the direct covariance up to a change of sign. Thirdly, the indicator covariance is not a connectivity index. A connectivity index tells us whether it is possible to find a path joining any two points in the facies and lying completely inside that facies, whereas the indicator covariance merely tells us whether to two end points are in the facies. It says nothing about paths joining the two points.

A facies F and its complement rarely have the same connectivity and yet their indicator covariances are identical. Figure 3.3 illustrates this point. The background is divided into two separate parts but the shaded facies is a single piece

Property 4: Three facies

Instead of two facies we now consider three facies labelled A, B and C, which together fill the whole space under study. See for example Fig. 3.4. At present we assume that the three random sets are stationary.

For these three facies:

$$\sigma_A(h) = -\sigma_{AB}(h) - \sigma_{AC}(h)$$
$$\sigma_B(h) = -\sigma_{BA}(h) - \sigma_{BC}(h)$$
$$\sigma_C(h) = -\sigma_{CA}(h) - \sigma_{CB}(h)$$
$$\sigma_{AB}(h) - \sigma_{BA}(h) = \sigma_{BC}(h) - \sigma_{CB}(h) = \sigma_{CA}(h) - \sigma_{AC}(h)$$

Proof. To prove this result, we note that any point x lies in one and only one of the three sets; that is:

$$1_A(x) + 1_B(x) + 1_C(x) = 1$$

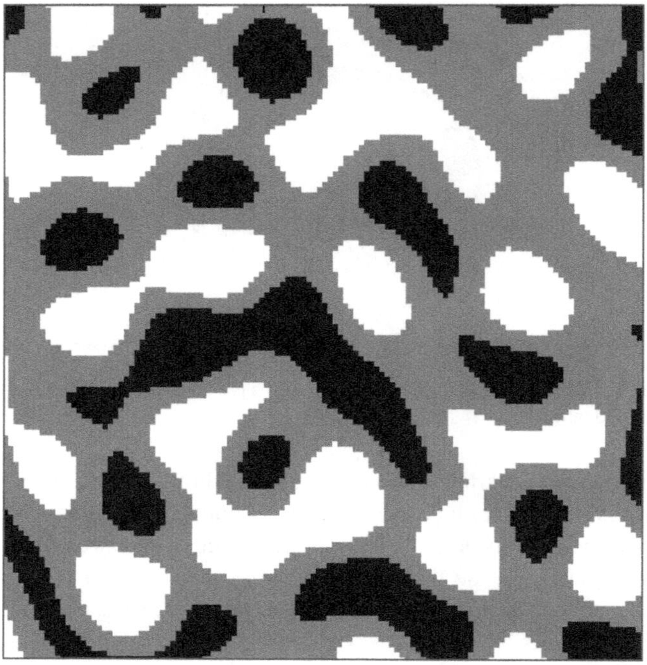

Fig. 3.4 Three facies A, B and C shown in *white*, *light grey* and *dark grey*

Consequently

$$P_A + P_B + P_C = 1$$

Hence

$$1_A(x) - P_A = -[1_B(x) - P_B + 1_C(x) - P_C]$$

We now substitute this into the second term of

$$\sigma_A(h) = E[1_A(x+h) - P_A][1_A(x) - P_A]$$

This gives

$$\sigma_A(h) = -E[1_A(x+h) - P_A][1_B(x) - P_B + 1_C(x) - P_C] = \sigma_{AB}(h) - \sigma_{AC}(h)$$

This proves the first three parts of property #3. Now to get the last part, we start from the definitions:

$$\sigma_{AB}(h) = E[1_A(x) - P_A]\,[1_B(x+h) - P_B]$$
$$\sigma_{BA}(h) = E[1_B(x) - P_B]\,[1_A(x+h) - P_A]$$

As before, we substitute the relation linking 1_A to $1_B + 1_C$ into these two, giving:

$$\sigma_{AB}(h) = -E[1_B(x) - P_B + 1_C(x) - P_C]\,[1_B(x+h) - P_B] = -\ \sigma_B(h) - \ \sigma_{CB}(h)$$
$$\sigma_{BA}(h) = -E[1_B(x) - P_B][1_B(x+h) - P_B + 1_C(x+h) - P_C] = -\ \sigma_B(h) - \sigma_{BC}(h)$$

Subtracting gives the result:

$$\sigma_{AB}(h) - \sigma_{BA}(h) = \ \sigma_{BC}(h) - \sigma_{CB}(h).$$

Property 5: Is it possible for two facies to be independent?

As A, B and C are a partition of the space, they cannot all be uncorrelated. But we can ask whether it is possible for two facies, A and B (say), to be uncorrelated. We are now going to show that this is not possible. If there were more than three facies, we could simply regroup all the other facies (except A and B) into a single facies C. So it suffices to prove the result for three arbitrary facies, A, B and C. Suppose that facies A and B are uncorrelated. Then

$$\sigma_{AB}(h) = E[1_A(x) - P_A]\ E[1_B(x+h) - P_B] = 0.$$

Similarly for $\sigma_{BA}(h)$. So

$$0 = \ \sigma_{AB}(h) - \sigma_{BA}(h),$$

hence

$$\sigma_{BC}(h) = \ \sigma_{CB}(h) \quad \text{and} \ \sigma_{CA}(h) = \ \sigma_{AC}(h).$$

Consequently

$$\sigma_A(h) = 0 - \sigma_{AC}(h)$$
$$\sigma_B(h) = 0 - \sigma_{BC}(h)$$
$$\sigma_C(h) = \ \sigma_{CA}(h) - \sigma_{CB}(h).$$

From this it is easy to show that

$$\sigma_C(h) = \ \sigma_A(h) + \ \sigma_B(h).$$

The next step is to substitute $h = 0$ into this equation. Since $\sigma(0)$ is equal to the variance, we get

$$P_C(1 - P_C) = P_A(1 - P) + P_B(1 - P_B)$$

and because $P_C = 1 - (P_A - P_B)$

$$P_A P_B = 0.$$

This means that one of the two facies A or B is not present, which contradicts the initial hypothesis. This proves that facies cannot be independent.

Property 6: One consequence of the triangle inequality

Matheron (1987) proved that all indicator variograms must satisfy the triangle inequality for any two distances, h_1 and h_2:

$$\gamma(h_1 + h_2) \leq \gamma(h_1) + \gamma(h_2)$$

This tells us about the behaviour of indicator variograms near the origin. Suppose that the behaviour was a power function near $h = 0$; say $\gamma(h) \approx |h|^\alpha$. What values are possible for the power α? Let $h_1 = h = h_2$. Then

$$\gamma(h_1 + h_2) \approx |2h|^\alpha \text{ and } \gamma(h_1) + \gamma(h_2) \approx 2|h|^\alpha$$

Consequently α must be less than or equal to 1. *This means that the gaussian variogram should not be used as a model for an indicator variogram.* Two examples of the types of inconsistencies that can arise when unsuitable variogram models are used for indicators, are given in exercises 2.3 and 2.4 at the end of the chapter. The point in including these examples is to show just how difficult it is to give a general characterisation for the variogram of an indicator random function, or in other words, of a random set. Matheron (1987, 1989, 1993) produced several papers on this subject. The third one gives results concerning groups of 3, 4 and 5 points and then makes a conjecture to generalise these. These results are relevant in "multi-point" geostatistics.

Need for a Mathematically Consistent Method

In the previous section we saw that although indicators are very simple themselves, their properties are far from intuitive. For example, most people believe that it would be possible for some of the facies or lithotypes to be independent of each other, until they have seen the proof to the contrary. Nor do they realise that power functions and the gaussian model are not suitable models for indicator variograms. Although we have proved that these two models are not acceptable, this does not mean that the others can be used. It merely means that no counter examples have yet found to them. Perhaps someone will one tomorrow. This raises a troubling question: which variogram models can be used for indicators. Or putting it more generally, Bochner's theorem provides a general characterisation for positive definite functions (i.e. for the covariances of gaussian random functions, and hence their variograms). Is there an equivalent theorem for random sets and indicator variograms? To the best of our knowledge there is not.

So we are faced with two options: make ad hoc choices about the indicator variograms with the risk of ending up with mathematical inconsistencies, or construct a mathematically consistent model. We prefer the second option. The way that we have chosen to do this is, as we saw in Chap. 1, by truncating simulations of gaussian random functions in a suitable way. The advantages of using underlying gaussian random functions are clear: firstly, the normal (gaussian) distribution has many nice statistical properties; secondly, there are many ways of simulating gaussian random functions. The good properties, from a theoretical point of view, of the normal distribution and gaussian random functions include

- In simple kriging the error is orthogonal to the estimate, so it is independent of it. This allows us to condition simulations by kriging.
- Any positive definite function can be used as the covariance model of a gaussian random function. This is not true for other distributions, either discrete or continuous. We have already seen this for indicators which are discrete variables. In a similar vein, Armstrong (1992) shows that the spherical variogram is not always compatible with a lognormal distribution.

Transition Probabilities

From experimental data, it is easy to compute the proportion of samples belonging to a particular facies. Although it is a little more complex, we can also calculate the conditional probability of going from one facies to another, or of staying in the same facies for points a certain distance apart. We are now going to express these transition probabilities in terms of the indicator covariance and the proportions.

First Type of Transition Probability

The first type of transition probability is just the probability of being in facies F_j at point $x + h$ knowing that point x is in facies F_i:

$$
\begin{aligned}
P(x + h \in F_j | x \in F_i) &= P\left[1_{F_j}(x + h) = 1 \,|1_{F_i}(x) = 1\right] \\
&= \frac{P\left[1_{F_j}(x + h) = 1 \;\&\; 1_{F_i}(x) = 1\right]}{P[1_{F_i}(x) = 1]} \\
&= \frac{E\left[1_{F_j}(x + h)\, 1_{F_i}(x)\right]}{E[1_{F_i}(x)]} = \frac{C_{ij}(x, x + h)}{p_{F_i}(x)}
\end{aligned}
$$

This is the probability of going from facies F_i to facies F_j, not being sure that we have a transition between x and x + h. This formula is also correct if j = i. If the facies are stationary, then this probability contains essentially the same information as the indicator variogram.

Example of the First Type of Transition Probability

To illustrate how to calculate this type of transition probability, we have selected four neighbouring drill-holes extending from one chronostratigraphic marker (N°11) down to the following one (N°10). Only two facies are present, A and B. Observations of the facies were made at 2 m intervals down the holes (Table 3.1).

As the drill-hole sections are of different lengths and as Marker N°10 is considered as the reference level (see Chap. 3 for more information on reference levels), the sections are given from this level upward rather than from marker N°11 downward, with blanks at the top. Our objective is to calculate the transition probabilities. By definition, these depend on the two points x and x + h. If the data are stationary horizontally (which is often the case), we can average along the rows. Similarly, if they are stationary vertically we can average in that direction but vertical stationarity is rare in sedimentary rock-types. Here as we only have four

Table 3.1 Facies observed at 2 m spacing going from marker N°11 down to marker N°10

Drill-hole N°92	Drill-hole N°58	Drill-hole N°39	Drill-hole N°84
A	–	–	–
B	–	–	–
A	A	–	–
B	A	–	–
A	A	B	–
A	A	A	–
A	A	B	–
B	A	B	–
B	A	A	–
B	B	A	–
B	A	A	–
B	A	B	B
B	A	B	B
B	B	B	A
B	B	B	A
B	B	B	B
B	A	B	A
B	A	A	B
B	B	B	A
A	A	B	A
B	A	B	B
A	B	B	B
B	B	B	B

Table 3.2 (a) Observed number of transitions for a distance of 2 m upward (out of a total of 71), (b) expected number of transitions assuming independence

(a)	Up to A	Up to B	(b)	Up to A	Up to B
From A	16	13	From A	13.7	17.5
From B	15	27	From B	17.5	22.3

Table 3.3 (a) Observed number of transitions for a distance of 10 m upward (out of a total of 55), (b) expected number of transitions assuming independence

(a)	Up to A	Up to B	(b)	Up to A	Up to B
From A	11	8	From A	10.7	13.5
From B	14	22	From B	13.5	17.3

drill-holes in this example, we will assume vertical and horizontal stationarity for simplicity.

Table 3.2a shows the numbers of each type of transition (i.e. A $\rightarrow$ A, A $\rightarrow$ B, B $\rightarrow$ A and B $\rightarrow$ B) for a distance of 2 m upward while Table 3.2b shows the number of transitions that would be expected, assuming a random distribution of facies. As 33 of the 75 observations are facies A, the Pr(A) = 0.44 and Pr(B) = 0.56. Consequently since there are 71 transitions each 2 m long, the expected number of transitions A $\rightarrow$ B or B $\rightarrow$ A is just $0.44 \times 0.56 \times 71$. In fact, the observed number of transitions A $\rightarrow$ A and B $\rightarrow$ B is higher than would be expected if the facies were arranged randomly, which was expected because of the sequential structure of sedimentary facies at this scale. Similarly, Table 3.3a shows the numbers of each type of transition for a distance of 10 m upward while Table 3.3b shows the corresponding number of transitions that would be expected, assuming independence.

One difference between the observed values in the previous tables is that Table 3.2 is almost symmetric around the diagonal but Table 3.3 is definitely not. The lower line representing sections starting with facies B contains 36 out of 55 observations (i.e. 65% rather than 56%). This indicates that facies B tends to be lower in the stratigraphic sequence than facies A; that is, this is evidence of non-stationarity. So averaging along the vertical direction is not a valid operation in this case. This problem will be encountered again when calculating variograms.

Second Type of Transition Probability

Here we consider the case where there are three or more facies. We are interested in the probability of being in facies F_j at point x + h knowing that we are in facies F_i at point x but not at x + h. This probability is:

$$
\begin{aligned}
&P(x+h \in F_j \,|x \in F_i \;\&\; x+h \notin F_i) \\
&\quad = P\left[1_{Fj}(x+h)=1\,|1_{Fi}(x)=1 \;\&\; 1_{Fi}(x+h)=0\right] \\
&\quad = \frac{P\left[1_{Fj}(x+h)=1 \,\&\, 1_{Fi}(x)=1 \,\&\, 1_{Fi}(x+h)=0\right]}{P[1_{Fi}(x)=1 \,\&\, 1_{Fi}(x+h)=0]} \\
&\quad = \frac{E\left[1_{Fj}(x+h)\,1_{Fi}(x)[1-1_{Fi}(x+h)]\right]}{E\{1_{Fi}(x)[1-1_{Fi}(x+h)]\}} \\
&\quad = \frac{E\left[1_{Fj}(x+h)\,1_{Fi}(x)\right]}{E\{1_{Fi}(x)[1-1_{Fi}(x+h)]\}} = \frac{C_{ij}(x,x+h)}{P_{Fi}(x)-C_{ii}(x,x+h)},
\end{aligned}
$$

where $C_{ij}(x, x+h)$ is the non centred covariance between $1_{F_i}(x)$ and $1_{F_j}(x+h)$. This is the probability of going from facies F_i to facies F_j, when we know that the facies at $x + h$ is not the same as at x. Clearly this probability cannot be computed if $j = i$.

Example

The second type of transition probabilities were calculated in the vertical direction for the image shown in Fig. 3.5, which was obtained by truncating a gaussian with an anisotropic gaussian variogram. The EW range was 50 units compared to 10 units in the NS direction. The graph shows the probability that the second point $(x + h)$ lies in facies 2 (or facies 3) given that the first point (x) is in facies 1 and that the second one is not in facies 1 (right). We see that for short distance up to about 4 units the only possible transition is from facies 1 to 2. As the distance increases the probability that the second point lies in facies 3 rises steadily to about 0.35.

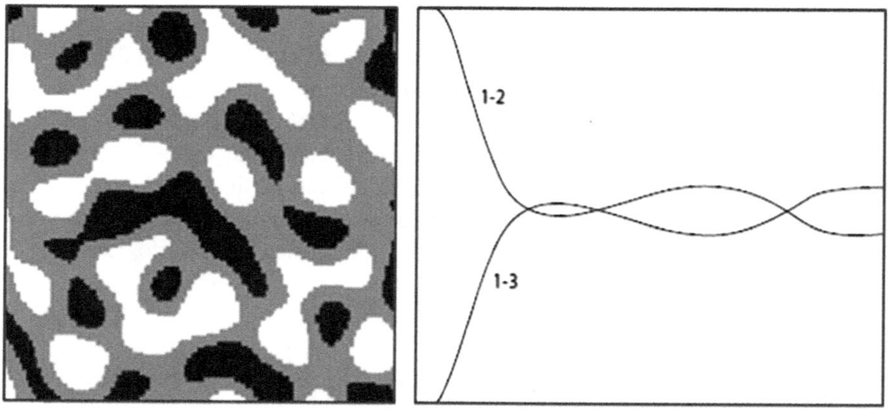

Fig. 3.5 Simulation of a gaussian random function with a gaussian variogram with an EW range of 50 units and a NS range of 10 units (*left*), and the probability of a point at position $x + h$ being in facies 2 (or facies 3, respectively) given that the first point at position x lies in facies 1 and that the second one does not, plotted as a function of the distance between the two points (*right*)

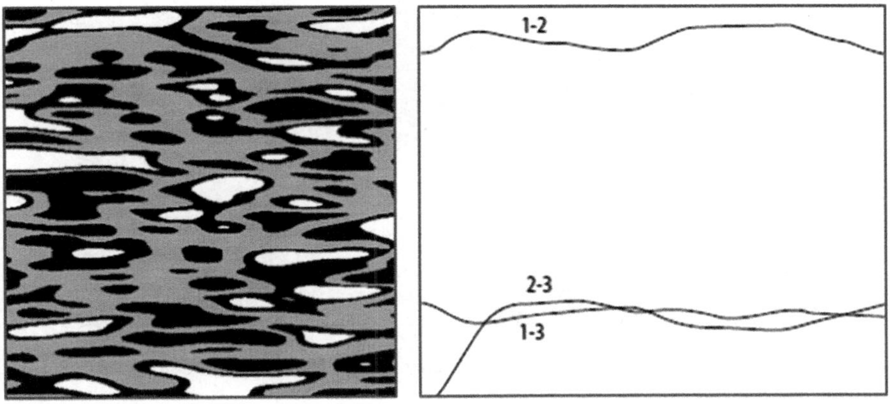

Fig. 3.6 Simulation of a gaussian random function with a gaussian variogram with an EW range of 50 units and a NS range of 10 units (*left*), and the probability of a point at position x + h being in facies 2 (or facies 3, respectively) given that the first point at position x lies in facies 1 and that the second one does not, plotted as a function of the distance between the two points (*right*)

As a contrast, the image shown in Fig. 3.6 was obtained by truncating the same gaussian random function using a different rock type rule. In the previous case, the rock type rule consisted of three intervals in the order: black, then grey and then white. Now there are four intervals: black, grey, black and white. The graph shows the corresponding transition probability. The essential difference between these two is that in the first case the black never touches the white whereas it does in the second case.

Exercises

The first two exercises are designed to familiarise readers with indicator random functions. The next two illustrate the types of inconsistencies that arise with unsuitable models for indicator covariances. The last exercise gives some rather interesting results on multi-point statistics.

Exercise 3.1

- Figure 3.7 shows 12 regularly spaced points in a 2D zone containing two facies (coloured grey and white). Let F denote the white facies. Write the values of the indicators of F and $\overline{F}$ in Table 3.4.
- Calculate the experimental mean and variance of the indicator for F and of its complement. Check that the means lie between 0 and 1, and that the variances are no greater than 0.25. Under what circumstances would the variances equal 0.25?
- Calculate the experimental variogram for facies F, and then for its complement for up to 3 lags in the EW direction. Plot these. Why are they the same? What shape do they have near the origin?

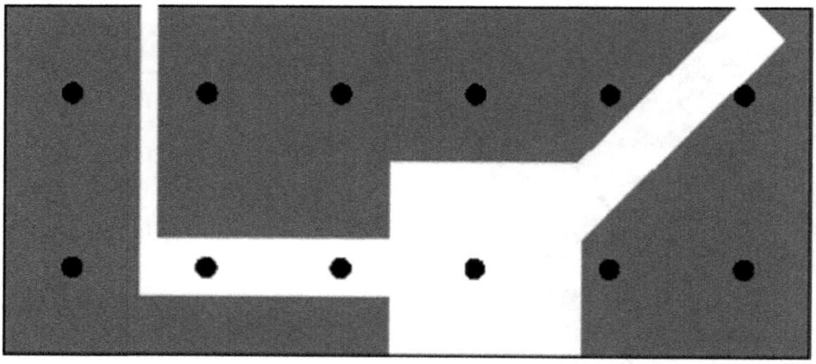

Fig. 3.7 Twelve samples are available in an area containing a facies F that is coloured in *white*. Its complement $\overline{F}$ is *shaded grey*

Table 3.4 Values of the indicator for the facies F and its complement

Point N°	1	2	3	4	5	6
$1_F(x)$						
$1_{\overline{F}}(x)$						
$1_F(x) + 1_{\overline{F}}(x)$						
Point N°	7	8	9	10	11	12
$1_F(x)$						
$1_{\overline{F}}(x)$						
$1_F(x) + 1_{\overline{F}}(x)$						

- The next step is to calculate the centred spatial covariances for F and its complement, then their cross covariances. Check that the direct covariances are identical and that the cross covariances are the negative of these.

Exercise 3.2

Assume that a facies F is stationary with probability $P_F = 0.6$ in the area under study. Show that the variance of the indicator $1_F(x)$ equals 0.24. Suppose that its variogram can be modelled as an exponential with a scale parameter equal "a".

- What is the value of its sill?
- Write down the equation for the variogram of its complement.
- Write down the equations for the centred covariance of the indicator for F and for its complement.
- Plot the non-centred covariances for these two indicators.

Exercise 3.3

Inappropriate models for indicator variograms. This example taken from Matheron (1987) and Armstrong (1992) highlights the types of inconsistencies that can arise when inappropriate models are chosen as indicator variograms. In this case we are going to show again that the gaussian variogram is unsuitable as a model for the variogram of the indicators.

Consider two points x_1 and x_3. Let x_2 be their midpoint. As each of the three indicators can take two values, 0 and 1, there are 8 possible combinations for the values of the three indicators. Let ω denote the probability that all three indicators take the value 1. That is,

$$\omega = \Pr[1_F(x_1) = 1_F(x_2) = 1_F(x_3) = 1]$$

As usual the non-centred covariance is:

$$C_{ij} = C_F(x_i, x_j) = E\left[1_F(x_i)1_F(x_j)\right] = \Pr\left[1_F(x_i) = 1_F(x_j) = 1\right]$$

Show that:

$$\begin{aligned}
\Pr[1_F(x_1) = 1_F(x_2) = 1 \ \& \ 1_F(x_3) = 0] &= C_{12} - \omega \\
\Pr[1_F(x_2) = 1_F(x_3) = 1 \ \& \ 1_F(x_1) = 0] &= C_{23} - \omega \\
\Pr[1_F(x_1) = 1_F(x_3) = 1 \ \& \ 1_F(x_2) = 0] &= C_{13} - \omega
\end{aligned}$$

Hint. Split the event $[1_F(x_1) = 1_F(x_3) = 1]$ into two mutually exclusive events according to the value of the third indicator e.g.

$$\begin{aligned}
\Pr[1_F(x_1) = 1_F(x_2) = 1] = \Pr[1_F(x_1) = 1_F(x_2) = 1_F(x_3) = 1] + \Pr[1_F(x_1) \\
= 1_F(x_2) = 1 \,\&\, 1_F(x_3) = 0]
\end{aligned}$$

Similarly show that

$$\begin{aligned}
\Pr[1_F(x_1) = 1 \ \& \ 1_F(x_2) = 0 \ \& \ 1_F(x_3) = 0] &= C_{11} - C_{12} - C_{13} + \omega \\
\Pr[1_F(x_1) = 0 \ \& \ 1_F(x_2) = 1 \ \& \ 1_F(x_3) = 0] &= C_{22} - C_{12} - C_{23} + \omega \\
\Pr[1_F(x_1) = 0 \ \& \ 1_F(x_2) = 0 \ \& \ 1_F(x_3) = 1] &= C_{33} - C_{13} - C_{23} + \omega
\end{aligned}$$

Hint. consider the event $[1_F(x_1) = 1 \ \& \ 1_F(x_2) = 1_F(x_3) = 0]$.

The contradiction becomes apparent if the three points are set 0.15 apart (compared to a scale parameter of 1) and if the probability that $1_{F_i}(x) = 1$ is 0.1.

$$C_{11} = C_{22} = C_{33} = 0.9$$

Calculate the value of the gaussian variogram for a distance of 0.15 and show that the following values are obtained for the noncentred covariance:

$$C_{12} = C_{23} = 0.1 \times 0.9 \times 0.494 = 0.0396 \text{ and } C_{13} = 0.1 \times 0.9 \times 0.478 = 0.043$$

Show that $0 \leq 0.043 - \ \omega$ since $\Pr[1_F(x_1) = 1_F(x_3) = 1 \ \& \ 1_F(x_2) = 0] \geq 0$.
In the same way, use the fact that

$$\Pr[1_F(x_1) = 0 \ \& \ 1_F(x_2) = 1 \ \& \ 1_F(x_3) = 0] = C_{22} - C_{13} + C_{23} - \ \omega \geq 0$$

to prove that $\omega \geq 0.82$

Since it is impossible for $\omega \leq 0.043$ and $\omega \geq 0.821$, the gaussian variogram model is incompatible with indicator data.

Exercise 3.4

Another unsuitable variogram model. At first one might be tempted to think that the gaussian variogram is not allowable because it is infinitely differentiable or because it is quadratic near the origin. To show that this is not the case, Matheron produced a construction for a variogram model

$$f(x) = [a^2 e^{-ax} - b(2a - b)e^{-bx}]$$

where $a < b < 2a$ and $\omega^2 = 2ab - b^2$.

This is licit for indicators in 1-D provided that $\omega < b$. However he also proved that if $\omega > b$, the triangle inequality is not respected. To see this, let $b = 0.1$, $\omega = 1$ and $h = 0.2$.

Exercise 3.5

Multi-point Statistics. Covariances and variograms are examples of two point statistics. This exercise gives some theoretical results on multi-point statistics. Part (a) shows that once one of the three point covariances is known, all the others can be expressed in terms of it and of lower order statistics. Let

$$C_{FF\bar{F}}(x, y, z) = E(1_F(x) 1_F(y) 1_{\bar{F}}(z))$$

Show that:

$$\begin{aligned} C_{FF\bar{F}}(x, y, z) &= -C_{FFF}(x, y, z) + C_{FF}(x, y) \\ C_{F\bar{F}\bar{F}}(x, y, z) &= C_{FFF}(x, y, z) - C_{FF}(x, y) - C_{FF}(x, z) + P_F(x) \\ C_{\bar{F}\bar{F}\bar{F}}(x, y, z) &= -C_{FFF}(x, y, z) + C_{FF}(x, y) + C_{FF}(x, z) + C_{FF}(y, z) + 1 \\ &\quad - P_F(x) - P_F(y) - P_F(z) \end{aligned}$$

By permuting the order of the coordinates show that

$$C_{F\bar{F}F}(x, y, z) = -C_{FFF}(x, y, z) + C_{FF}(x, z)$$

Deduce similar expressions for the remaining three point statistics. (Exercise 2.3 is based on these.) We can now extend this result to higher order multi-point statistics. First we need to define the notation. Let

$$C_{FF}(x_1, x_2) = C_{F^2} \text{ and } C_{FFF}(x_1, x_2, x_3) = C_{F^3}$$

More generally, let the covariance for the case where the first (p−k) points belong to the facies F and the remaining k points to its complement be

$$C_{\underbrace{F...F}_{(p-k)}\underbrace{\overline{F}...\overline{F}}_{k}}(x_1, ...x_p) = C_{F^{p-k}\overline{F}^k}(x_1, ...x_p)$$

Show that for any two integers k and p

$$C_{F^{p-k},\overline{F}^k}(x_1, x_2, ..., x_p) = (-1)^k C_{F^p}(x_1, x_2, ..., x_p) + \Phi\left[(x_1, x_2, ..., x_p)\right]$$

The function Φ involves only lower order covariances $C_{F^{p-i}}$ and proportions. This can be deduced by expanding the appropriate product. For example, to find $C_{F\overline{F}\overline{F}}(x, y, z)$, we expand

$$F(x)[1 - F(y)][1 - F(z)] = F(x)F(y)F(z) - F(x)F(y) - F(x)F(z) + F(x)$$

This gives

$$C_{F\overline{F}\overline{F}}(x, y, z) = C_{FFF}(x, y, z) - C_{FF}(x, y) - C_{FF}(x, z) + P_F(x).$$

By permuting the coordinates we can show that any n-point statistics can be expressed using the basic n-point covariance $C_{F^p}(x_1, x_2, ..., x_p)$ and terms involving lower order covariances.

Exercise 3.6

Erosion model. Consider two models with two facies where A and D are the main facies and $\overline{A}, \overline{D}$ their complements. Then form a model with three facies, A, B and C, by letting A erode D and its complement.

- Write the indicators of A, B, C using the indicators of A, and D.
- Use the previous results to compute the indicator covariances of A, B, C for the case when the facies A is independent of D.

Exercise 3.7

Porous medium. Consider a porous media. Let A be the random set describing the pores. The porosity of support V is the mean value of the indicator A on the volume V centred at the point x. Compute the analytical form of

- The covariance of the porosity of support V.
- The covariance of another indicator F with the porosity of support V.

Chapter 4
Proportions

The simplest quantity that can be measured from an image showing different facies or lithotypes is the percentage of space taken up by each facies. So the first step in a plurigaussian study is to calculate the proportions from the experimental data. In the stationary case, we only need calculate the percentage of all the data that belong to each facies to get the proportions. However in most practical cases, the geology is more complicated. For example, petroleum reservoirs are not stationary in the vertical direction because of cyclic changes during their deposition. Vertical proportion curves were designed to quantify these changes. We will show how to compute and interpret these curves. In more complicated cases the proportions vary laterally as well. In that case we build up a 3D matrix of proportion curves. Although plurigaussian simulations were first developed for the oil industry, they are now used in mining as well. Proportion curves and 3D proportion matrices are also used there to estimate the proportions of each rock type.

The first section in this chapter describes vertical proportion curves for the simplest case where the reservoir or deposit is stationary horizontally. These curves summarise the vertical variability in the proportions. After that we treat the more general case of non-stationarity where we use 3D proportion matrices. One key factor when calculating vertical proportion curves (VPC, for short) is the choice of the reference level. An example is presented to illustrate the impact of inappropriate choices on VPC.

How to Calculate Vertical Proportion Curves

Vertical proportion curves first proposed by Matheron et al. (1987) are a simple tool for quantifying the evolution in the amount of each facies or lithotype present as a function of depth. They are computed along lines parallel to the chosen reference level (generally a chrono-stratigraphic marker). The results are presented as a graph showing the proportion of each facies at each level. We illustrate this procedure using a simple example.

Figure 4.1 shows five fictive wells each containing five core sections of equal length. Three lithotypes which we call sandstone, shaly-sandstone and shale, have

M. Armstrong et al., *Plurigaussian Simulations in Geosciences*,
DOI 10.1007/978-3-642-19607-2_4, © Springer-Verlag Berlin Heidelberg 2011

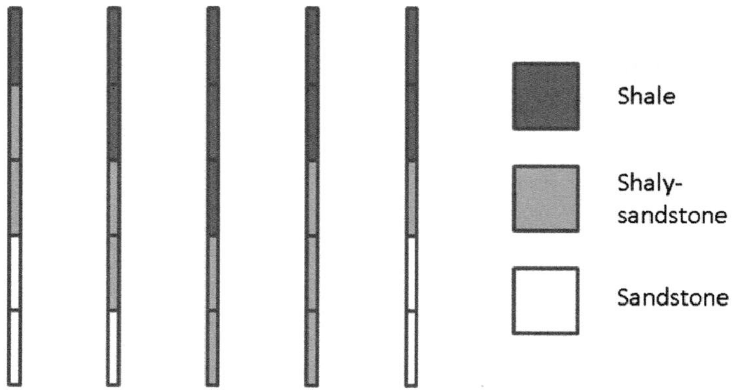

Fig. 4.1 Five drill-holes each containing five core sections of equal length

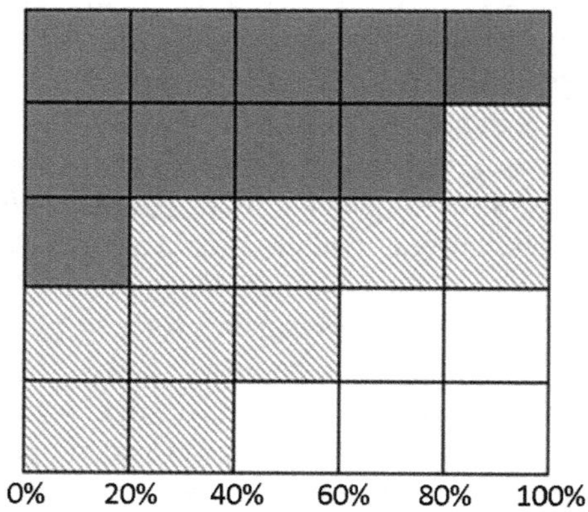

Fig. 4.2 Vertical Proportion Curves showing the proportion of each facies per level

been recorded. Looking at this figure, we see that there is more shale at the top of the wells than at the bottom. In fact, all five core sections in the top row are shale (i.e. 100% shale). In the second row there is 80% shale and 20% shaly-sandstone. At the bottom there is 40% shaly-sandstone and 60% sandstone, but no shale.

These proportions are presented graphically in Fig. 4.2. The order that the facies are arranged in is important. It must reflect the evolution seen in the geology. In this case the interpretation is simple: the grain size at the top of the wells is much smaller than at the bottom. This indicates sedimentological evolution from a high energy medium at the bottom to a low energy one at the top.

Example 1: The Ravenscar Sequence

The Ravenscar sequence which outcrops in cliff faces near Scarborough in North Yorkshire, consists of about 200 m of siliclastic sediments dating from the Middle

Jurassic era. The Cleveland basin to which it belongs is a progradation of a deltaic system with marine influences. Figure 4.3 (taken from Beucher et al. 2006) shows a conceptual model of it.

The Ravenscar sequence consists of seven units. From the top down, these are Scarborough, Gristhorpe, Millepore, Sycarham, Cloughton, Ellerbeck and Saltwick. Each had a distinctly different depositional environment. For example, the Saltwick unit was laid down in a continental environment whereas the Ellerbeck unit is characterised by steady changes in the sea level. Figures 4.4 and 4.5 (from Beucher et al. 2006) show the vertical proportion curves for these two units. The three main channel producing episodes are clearly visible in the VPC for the Saltwick unit. In contrast to this, the VPC for the Ellerbeck unit shows regular changes from an open

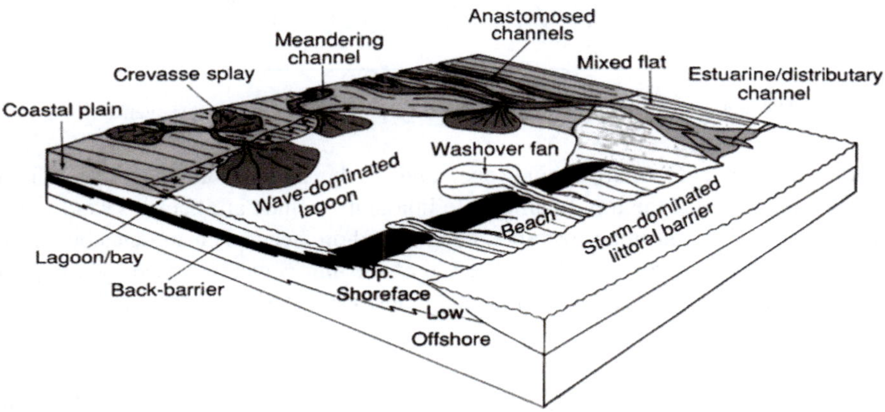

Fig. 4.3 Conceptual model of the deposition

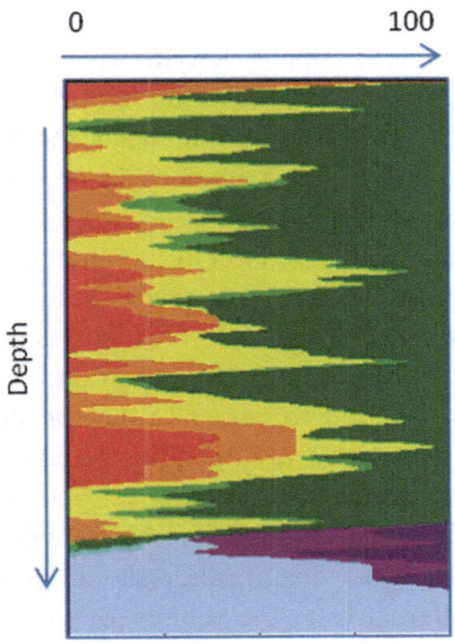

Fig. 4.4 Vertical proportion curves for the Saltwick unit of the Ravenscar sequence

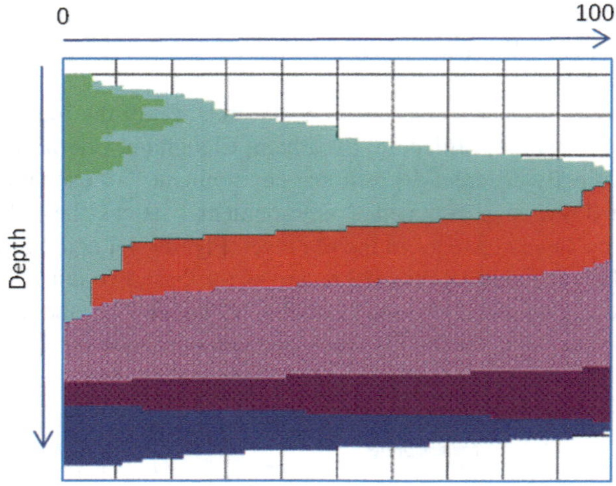

Fig. 4.5 Vertical proportion curves for the Ellerbeck unit of the Ravenscar sequence

marine environment (marine mudstones and marls) at the bottom of the unit to continental facies at the top. Intermediate lithounits include argillaceous sandstones and transgressive sandstone corresponding to the shoreface and clean sandstone from the foreshore and upper shoreface. The top of the unit is characterised by a lagoon environment with a floodplain, continental mudstone and argillaceous sands.

Example 2: Facies with Contrasting Anisotropies in a Gold Deposit

The Lupin mine in Northwest Canada is a good example of a stratiform banded iron formation (BIF) hosted gold deposit. Some of the gold is uniformly disseminated in thin, laterally continuous units of sulphur-rich BIF while the rest is contained in steeply inclined quartz veins that have overprinted the BIF. See Kerswill et al. (1996) for more information. The two gold-bearing lithofacies and the barren background are present in the schematic vertical cross-section of the orebody shown in Fig. 4.6 from Roth et al. (1998).

Figure 4.7 shows the experimental VPC computed from Fig. 4.6 with the proportion of quartz vein in dark grey and that of sulphide BIF in light grey. The proportion of quartz vein changes gradually, reaching a maximum near the middle of the cross-section. In contrast to this, the proportion of sulphide BIF is erratic.

Before this VPC can be used to calculate the thresholds we have to decide whether these peaks are representative of genuine geological features in the deposit or whether they are due to statistical fluctuations. Looking back at Fig. 4.4, we see that same problem arose with the VPC for the Saltwick unit of the Ravenscar sequence. If the peaks are a real feature they should be modelled. Otherwise we should smooth them. Alternatively, the reference level has to be changed. This decision can only be made after discussion between the geologist or the engineer

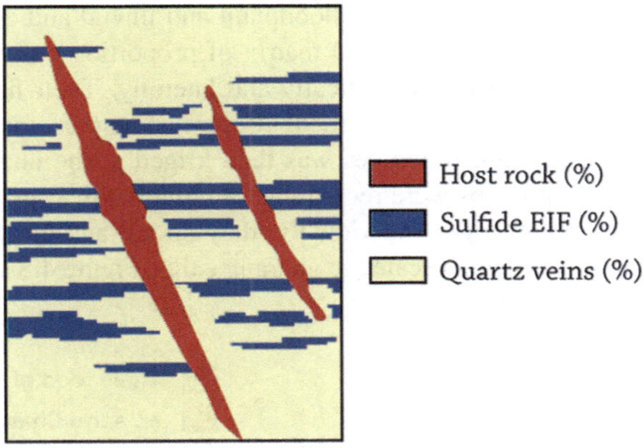

Fig. 4.6 Schematic vertical cross-section of the Lupin orebody showing two mineralised litho-facies with contrasting anisotropies on a waste background

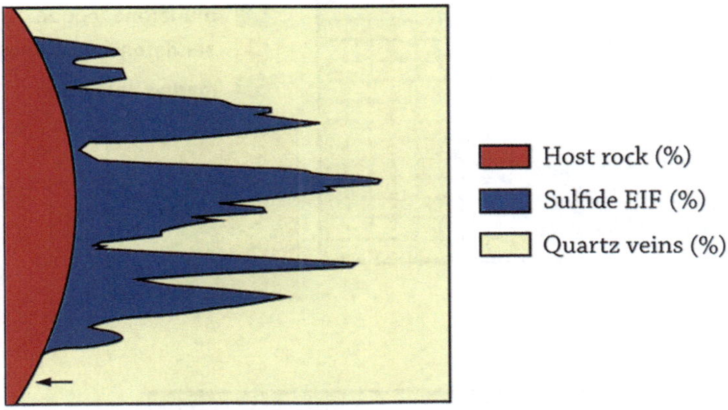

Fig. 4.7 The corresponding vertical proportion curves

and the geostatistician. In both of these cases it was decided that the peaks represented real features of the deposit/reservoir that should be reproduced in the simulations. So the thresholds were set to generate more BIF (or more channels as the case may be) at the corresponding depths.

Horizontal Non-stationarity

The vertical proportion curves presented in the previous section were computed assuming that the deposit/reservoir was at least horizontally stationary. In many cases, this is not true. The proportions vary laterally from one area to another. The Millpore and Gristhorpe units of the Ravescar sequence in Yorkshire are a typical example of this. The depositional environment varied from marine to littoral deposits. It is characterised by marine mudstone, wash-over argillaceous sandstones through to deltaic environment with a floodplain and fluvial and deltaic channels. Beucher et al. (2006) constructed a 3D matrix of proportion curves to model the evolution in the proportions both vertically and laterally. Their first steps was to group wells locally and compute vertical proportion curves for each set. The percentage of each of eight lithotypes was then kriged at the nodes of a regular grid and the kriged estimates were recombined to give vertical proportion curves. As the proportions were kriged individually, they do not add up to 100% at each grid node. So the values were rescaled to guarantee this. Figure 4.8 shows nine cells from the 3D proportion matrix.

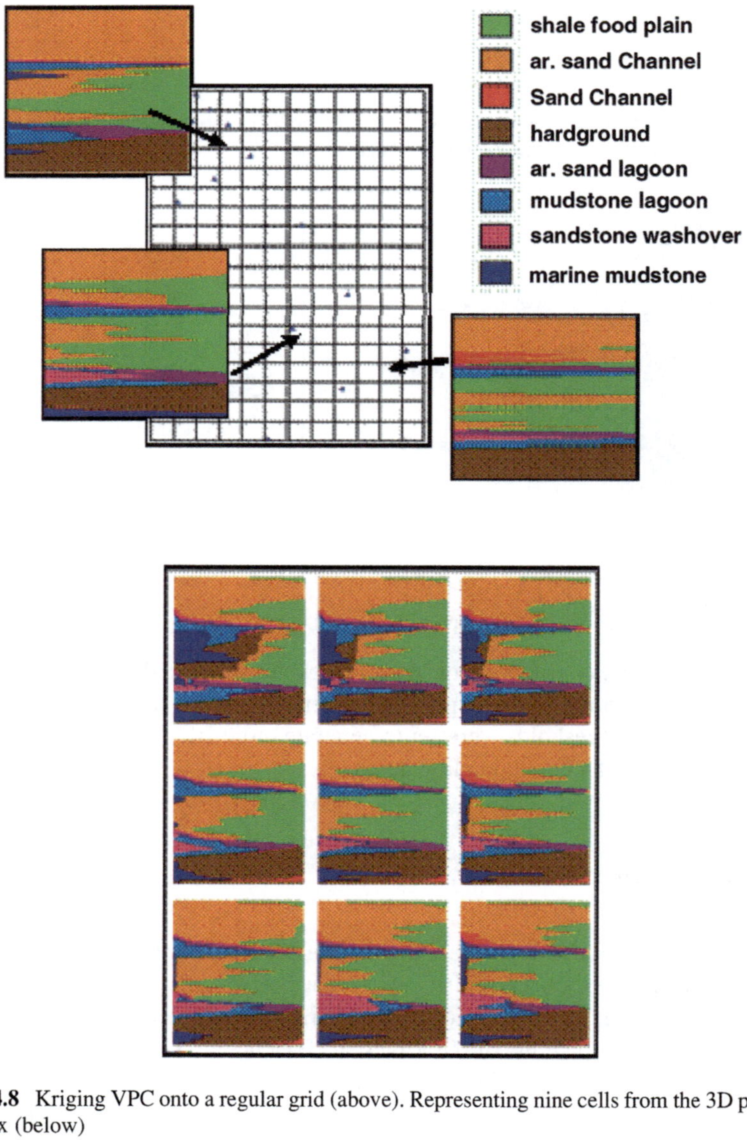

Fig. 4.8 Kriging VPC onto a regular grid (above). Representing nine cells from the 3D proportion matrix (below)

Having to rescale the proportions to make them sum to 100% is not entirely satisfactory. One could envisage using cokriging, but this would result in difficulties setting up the structural model (i.e. the variograms and the cross-variograms). A more fundamental reason for using a simple technique like kriging is that it is then easier to incorporate additional information, for example from seismic data. Moulière et al. 1997; Moulière 1998 considered a case where a 2D seismic attribute gave the cumulative thickness of one lithotype (shaly sandstone) for each unit in the formation. So it was important to incorporate this constraint into the proportion curves. She proposed two methods for doing this: one based on cokriging and the other using kriging. She showed that the simulations obtained after taking account of the available seismic information were much more realistic than those based on data from the limited number of wells that were available.

Lateral non-stationarity was also encountered by Doligez et al. (1994a, b) in their study of the Cajigar 2 succession. As the proportions changed from one area to another, they constructed a matrix of lithofacies proportions on a 300 m × 300 m grid. Each horizontal proportion curve was obtained by averaging the percentages of each lithofacies in the three levels above and below, as well as the level itself. This has the advantage of maintaining the vertical correlations between successive levels, as well as smoothing the results. Whereas in vertical proportion curves, the proportions are plotted horizontally as a function of the vertical height, the horizontal proportion curves show the proportions on the Y-axis with horizontal coordinate along the X-axis.

Non-stationarity can also arise when studying mining deposits. Compared with petroleum data, far more drill holes are usually available but there is often no chronostratigraphic marker to help interpret the geology. Betzhold and Roth (2000) studied the Mantos Blancos copper orebody which is located 45 km northeast of the city of Antofagasta in the north of Chile.

Their objective in simulating the orebody was to improve the ore homogenisation procedure by providing mine-planning engineers with more accurate images of the key mineralogical units. The rocks at Mantos Blancos come from a volcanic sequence consisting of andesitic flows and flow breccia at the top, then flow-breccia and flows of porphyritic dacite, and at the bottom, flows of augen (quartz-eye) dacite.

The sequence dips to the southwest at 10–20° in the mine area. The mineralisation forms an irregular blanket ranging from 100 to 200 m thick. As different mineralogical ensembles with similar average grades do not increase the grade variability, they can effectively be grouped together and treated as one new unit.

This led Betzhold and Roth to define three ore type classifications:

- High grade ore comprising chalcocite-bornite ore
- Low grade ore comprising chalcopyrite and chalcopyrite-pyrite ores
- Waste rocks that are not sent to the flotation plant. These include the oxide ore (atacamite-malachite), pyrite and the barren rock that is generally found beneath the orebody

Over the whole field, there is almost 10% of high-grade ore and about 45% of both low grade and non-copper sulphide material but these proportions are not constant in space. For example, there is much lower grade material near the top of the orebody. The proportions have to be modelled as a function of the point considered in the field.

Proportion curves were computed by a simple moving average procedure that was refined and complemented manually to incorporate knowledge about the geological characteristics of the site. Figure 4.9 shows two sets of proportion curves from a 2D test zone containing just over 900 samples. The horizontal axes are the east-west direction (on the left) and the vertical direction (on the right). In the top diagram, the horizontal proportion curve was calculated close to the surface, and the vertical proportion curve on the Western side of the field. In the lower diagram

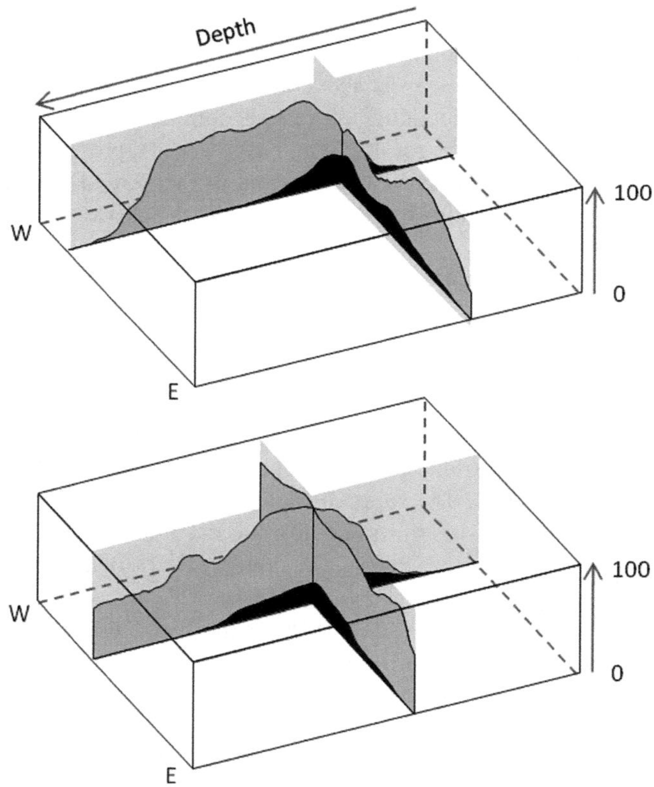

Fig. 4.9 Two sets of proportion curves, an HPC and a VPC, calculated using just over 900 samples. The horizontal axes are the east-west direction (on the *left*) and the vertical direction (on the *right*). In the top diagram, the HPC was calculated close to the surface, whereas the VPC came from the Western side of the field. In the lower diagram both proportion curves were computed closer to the centre of the field. The proportions are shown as a coloured surface on the vertical axis with high grade ore in black, low grade ore in *middle grey* and poor material in *light grey*

the proportion curves were computed closer to the centre of the field. The proportions of the three ore types are shown as a coloured surface on the vertical axis: high grade ore in black, low ore in middle grey and poor grade in light grey. The sum of the facies proportions is always equal to 100%. It is interesting to see how much the proportions of low and poor facies can vary over relatively small distances. In contrast the proportion of high grade ore seems more regular.

Choosing the Reference Level

The shape of the vertical proportion curves and the resulting simulations depend on the choice of the reference level. For oil reservoirs this is a specific geological marker which is used to restore the geometry of the reservoir at the time of deposition. This level must have been horizontal during sedimentation, and, should, if possible, correspond to a time line. That is, it should be a chronostratigraphic marker not an erosional unconformity. The reservoir is then flattened using this as the reference level.

Common choices for the reference level are the top of the unit or the bottom. Alternatively, a proportional grid could have been used to take account of differential subsidence. Figure 4.10 illustrates two of these three possibilities. Different reference levels can be used for each reservoir unit, or one serve for several of them.

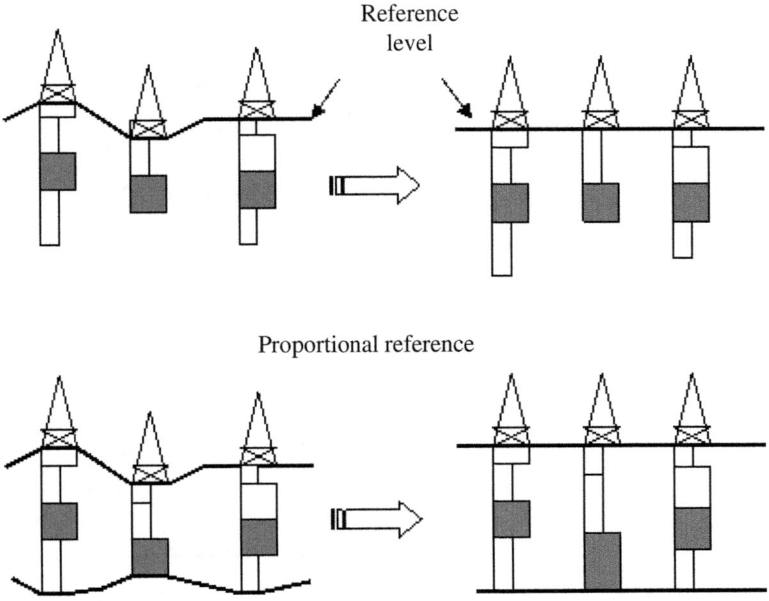

Fig. 4.10 Two ways of flattening a reservoir, (**a**) using the top as reference level, (**b**) proportionately to account for differential subsidence

To illustrate how important this choice is, Figure 4.11 illustrates the impact of three possible reference levels on the vertical proportion curve and then on the simulations. The Ellerbeck unit of the Ravenscar sequence which was described earlier in the chapter, was used as an example. The correct level is a flooding surface inside the Ellerbeck unit near the bottom. The two incorrect levels she considered were the base of the Ravenscar sequence and its top (denoted by R1 and R2 respectively). The base is a chronostratigraphic marker, the top is an unconformity.

Figure 4.12 shows the VPC for the three different choices. The best one (denoted by OK) shows a steady gradation in the depositional environment. Starting from the bottom these are:

- Mudstone coming from an open marine environment
- Argillaceous sandstone from the low to mid shore
- Transgressive sandstone
- Clean sandstone from the foreshore
- A mixture of argillaceous sandstone and mudstone from a lagoon environment
- Continental mudstone from a flood plain environment at the top

The second VPC correspond to reference level R1 which is not the correct one. But its irregularities could have been due to cycles within the depositional cycles. The third VPC corresponding to reference level R2 is obviously incorrect. Figure 4.13 shows one simulation of the unit obtained using the correct VPC which reproduces the type of continuity that would be expected at this scale, in this type of depositional environment. The lithotypes in the simulation corresponding to R1 are much less continuous

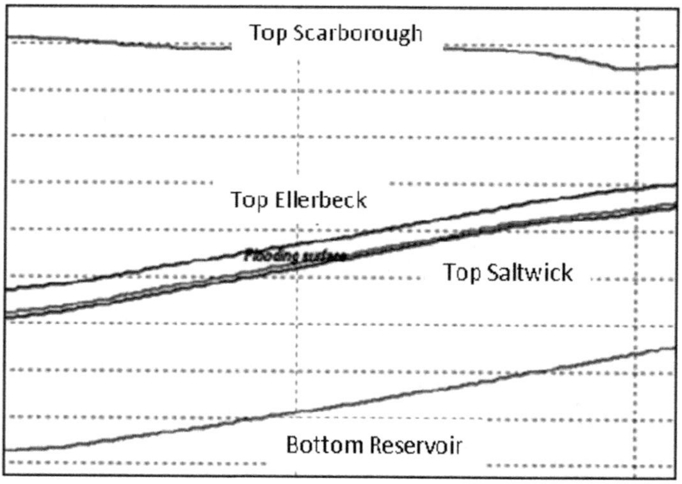

Fig. 4.11 Three possible choices for the reference level; the flooding surface inside the Ellerbeck formation (the correct choice), the bottom of the reservoir (R1) and the top (R2)

Fig. 4.12 Vertical proportion curve corresponding to the different reference levels: correct (top) – reference level R1 (middle) – reference level R2 (bottom)

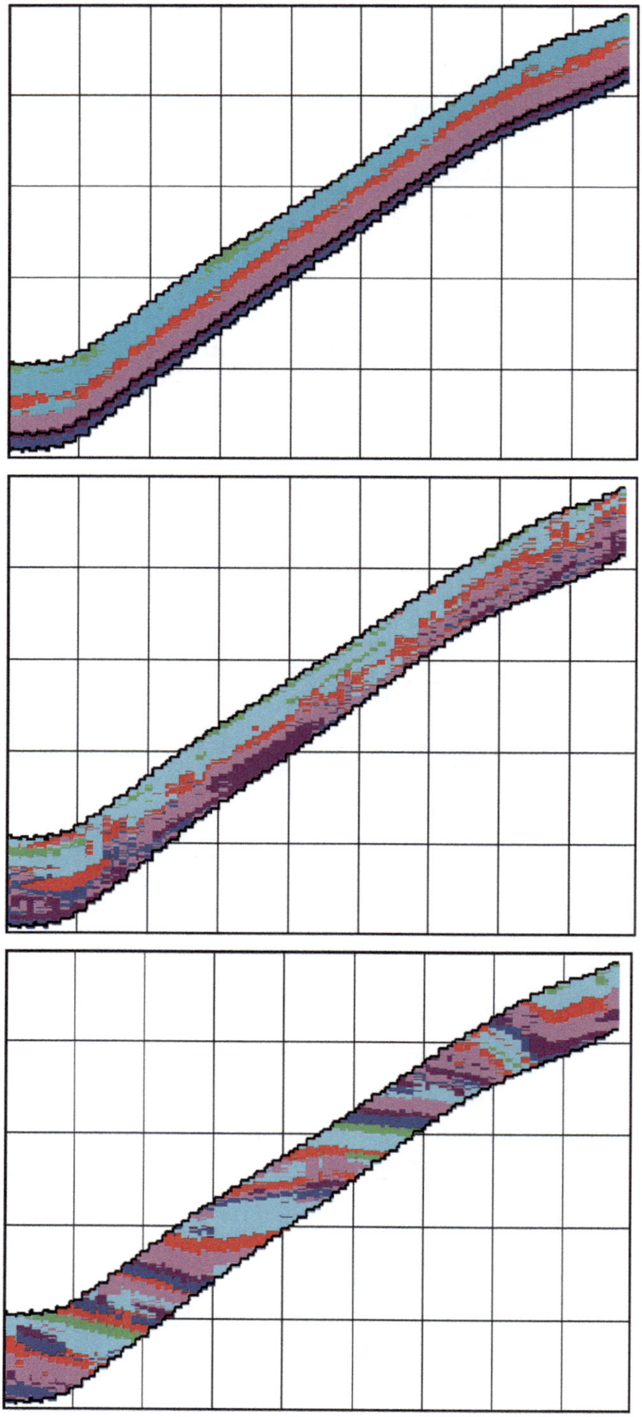

Fig. 4.13 One simulation of the unit obtained using different reference levels: correct (top) – reference level R1 (middle) – reference level R2 (bottom)

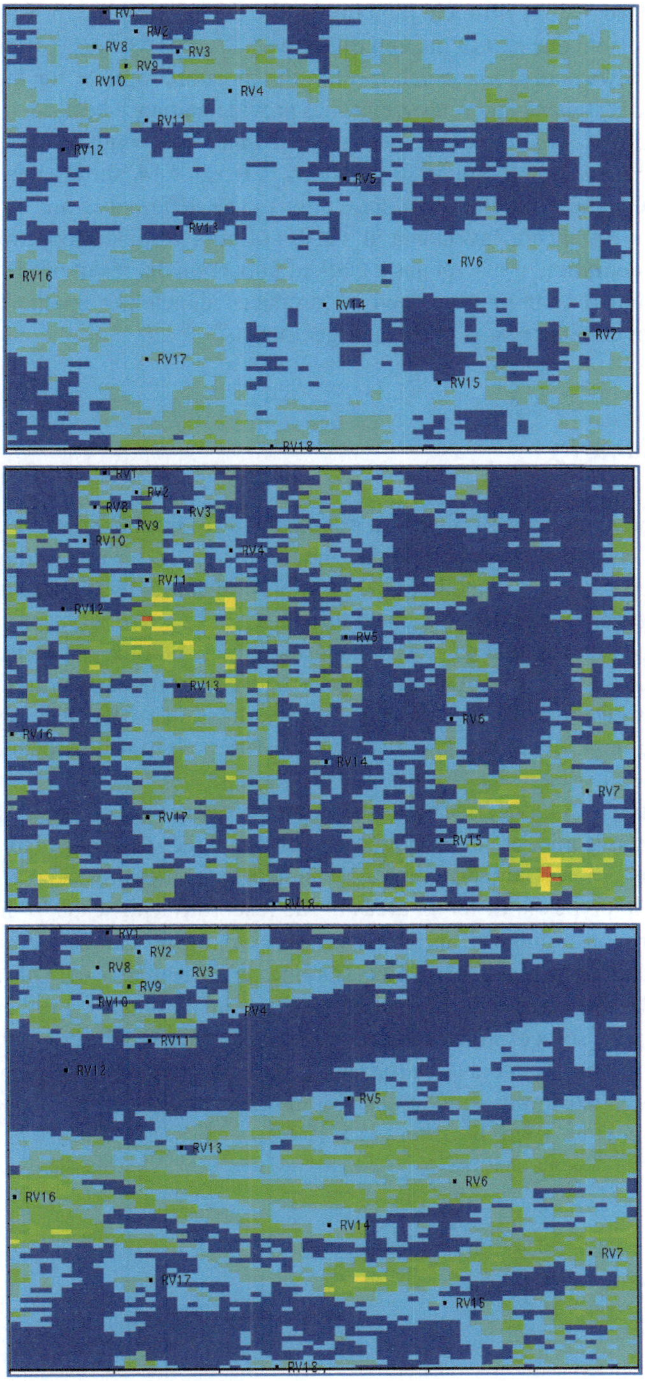

Fig. 4.14 Accumulated sand thickness obtained using the different reference levels: correct (top) – reference level R1 (middle) – reference level R2 (bottom)

and would lead to quite different fluid flows. The third simulation for R2 suggests no connectivity from one part of the unit to another.

It is not always obvious which level is the most appropriate for the reference. When in doubt, seismic data can sometimes provide guidance because it can give an indication of the cumulative thickness of sandstone. This could be compared to the cumulative thickness found in simulations. Figure 4.14 shows the cumulative sandstone thickness for the three simulations. The differences are quite marked. Volpi et al. (1997) and Ravenne et al. (2002) discuss vertical proportion curves in detail. The definition of a level has a marked impact on the result of the simulation. Onlap or toplap configurations can be produced depending on the choice of a reference level with regards to the unit geometry.

Non-stationarity

From a geological point of a view, a non-stationarity is characterised by a significant lateral change in the lithotype distribution in a reservoir unit, within the study area. It often shows up in the horizontal proportion curves as significant variations in the lithotype proportions in a given direction.

As geological phenomena always have some non-stationary aspects, depending on the scale of investigation, the choice between stationary or non-stationary models is subjective. It implies firstly a classification of the heterogeneity and then setting up a hierarchy. The distribution of the relevant heterogeneity will be then analysed to determine if it is stationary or not.

Non-stationarity concerns both object and sequence-based models. In object based models, it implies lateral variations of the object frequency. In sequence – based models, non stationarity can be handled by computing several proportion curves for lithotypes in different areas of the reservoir. This leads to the construction of a proportion matrix, in which vertical proportion curves are computed in each cell of a grid (Beucher et al. 1993; Doligez et al. 1999a, b, c). This approach is very flexible and gives realistic lateral variation of lithofacies. Furthermore, it is easy to integrate soft constraints such as an external drift into the computation of the proportion matrix. Geological information derived from seismic campaigns can then be integrated in the simulations (Moulière et al. 1997; Fournier and Derain 1997; Johann et al. 1996; Beucher et al. 1999).

Chapter 5
Truncation and Thresholds

Basic Principle in the Truncated Gaussian Method

In truncated gaussian simulations, the lithofacies are not simulated directly: a stationary gaussian random function is simulated first, and is then transformed into the lithofacies variable by truncation. For example, if we want to simulate two lithofacies, F_1 and F_2, a very intuitive way to transform the simulated gaussian variable into facies values is to say "if the numerical value of the simulated gaussian is lower than the number t_1, we obtain the first facies F_1; otherwise, we obtain the second facies F_2". The value t_1 is called a threshold.

Defining the Thresholds

Let x be any point in the simulated domain, let $1_{F1}(x)$ and $1_{F2}(x)$ be the indicators of the lithofacies F_1 and F_2 and let Z(x) be the simulated gaussian function at point x. The transformation used is described mathematically by:

$$1_{F1}(x) = 1 \quad \Leftrightarrow \quad -\infty \leq Z(x) < t_1.$$

When we want to simulate a lithofacies variable which can take more than two values, we have to define more than one threshold (N$-$1 thresholds for N possible lithofacies values). The ith facies F_i is defined by:

$$x \in F_i \quad \Leftrightarrow \quad 1_{Fi}(x) = 1 \Leftrightarrow t_{i-1} \leq Z(x) < t_i.$$

The thresholds are in increasing order

$$t_1 \leq t_2 \leq \ldots \leq t_{i-1} \leq t_i \leq t_{i+1} \leq \ldots \leq t_{N-1}.$$

And we can obviously have one and only one facies value at each point.

M. Armstrong et al., *Plurigaussian Simulations in Geosciences*,
DOI 10.1007/978-3-642-19607-2_5, © Springer-Verlag Berlin Heidelberg 2011

Transitions Between Facies

Provided that the nugget effect in the variogram is zero, the simulated gaussian function is continuous. We are going to study the simulated values as we go from point x to point y.

For any intermediate value between z(x) and z(y), there must be at least 1 point where z takes that value. This is true for every path going from x to y, no matter how complicated it is (Fig. 5.1).

This has important consequences. Suppose that the points x and y belong to different facies, respectively F_i and F_j, where $i < j$. Then we will find all the intermediate facies F_k ($i < k < j$) on all paths joining the points x and y (provided that the thresholds do not vary in space). Graphically, this corresponds to a

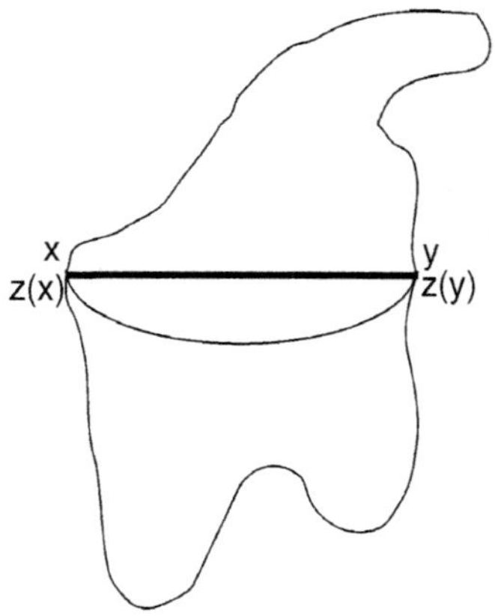

Fig. 5.1 Examples of possible paths

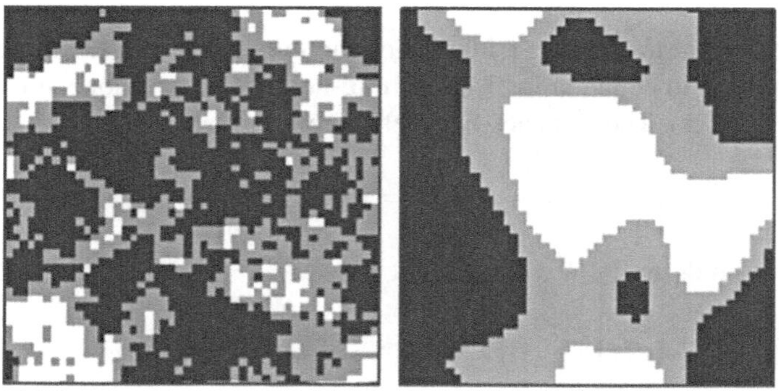

Fig. 5.2 Characteristic patterns generated using an exponential variogram (*left*) and a gaussian variogram (*right*)

characteristic pattern of concentric shapes (Fig. 5.2). In both cases, if we go from a point in the white facies to one in the dark grey, we have to cross the light grey. So the contacts that are possible using the truncated gaussian model are defined by the ordering of the thresholds, i.e., by the ordering of the facies.

Figure 5.3 shows three simulations obtained with different choices of facies ordering. In the top figure, the light grey always occurs between white and dark

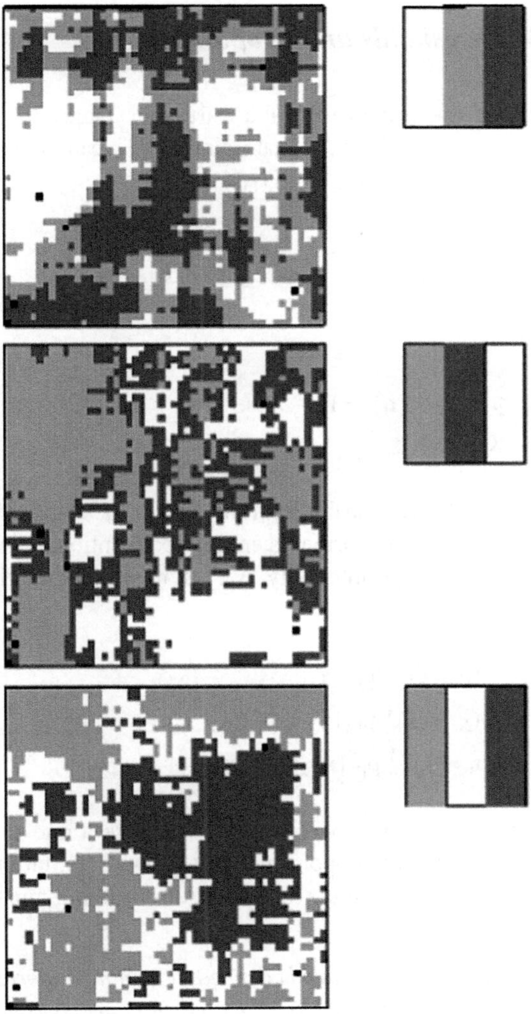

Fig. 5.3 Three truncated gaussian simulations with the facies ordered in different ways. At the top, the *light grey* always occurs between *white and dark grey*. In the middle, the *dark grey* forms the buffer between the other two colours whereas at the bottom it is the *white*. This relationship is summarised in the rocktype rules to the right of each figure

grey. In the middle figure, the dark grey forms the buffer between the other two colours whereas in the bottom figure it is the white. This relationship is summarised in the rocktype rules to the right of each figure. The discretisation of the space can attenuate this effect, especially when the properties of the gaussian variable allow sharp (but nevertheless continuous) variations.

Variation of the thresholds in space can also attenuate this effect, especially when some facies disappear locally.

Link Between Thresholds and Proportions

Earlier we saw that the proportion of a particular facies F_i at point x is the probability of having this facies F_i at that point x. This can be written as:

P_{Fi} = P(facies at point x = F_i) = $E[1_{Fi}(x)]$.

As we have

$$1_{Fi}(x) = 1 \Leftrightarrow t_{i-1} \leq Z(x) < t_i,$$

we can also write:

$$p_{Fi}(x) = P(t_{i-1} \leq Z(x) < t_i) = P(-\infty < Z(x) < t_i) - P(-\infty < Z(x) < t_{i-1})$$
$$p_{Fi}(x) = G(t_i) - G(t_{i-1}),$$

where G(t) is the cumulative distribution function for the standard normal distribution N(0,1). (The choice of the normal is entirely conventional). As the proportions of each facies are known experimentally, we just invert this relationship to deduce the thresholds:

$$t_1 = G^{-1}[p_{F1}(x)]$$
$$t_2 = G^{-1}[p_{F1}(x) + p_{F2}(x)]$$
$$t_i = G^{-1}[p_{F1}(x) + p_{F2}(x) + \cdots + p_{Fi}(x)].$$

Once the facies ordering has been chosen, we have a one to one relationship between the thresholds and the proportions. This shows a limitation of the truncated gaussian method: as the model orders the facies, this ordering must be realistic from a geological point of view.

Remark 5.1. Once the facies sequence is defined, it is of no importance whether they are numbered in an increasing order or a decreasing one. By replacing Z with −Z, both cases are equivalent.

Non Stationary Facies

If the facies are stationary, the thresholds are constant in space; otherwise they vary. The example of how to calculate vertical proportion curves given in Chap. 3 is a typical case of vertical non stationarity. Table 5.1 lists the proportions of each facies in each level for that example and Table 5.2 gives the corresponding thresholds.

The first and last thresholds are always equal to $-\infty$ and ∞ respectively, and they are not taken into account in the N$-$1 thresholds (if we consider them, there are N+1 thresholds).

In non stationary cases, facies sometimes disappear locally. For example, if the ith facies is not present, then $t_{i-1} = t_i$ as occurred in Table 5.2.

Idea Behind the Plurigaussian Method

With the plurigaussian method, we simulate *several* gaussian functions instead of one. A transformation is made on the whole set of gaussian functions to obtain the facies value. We will illustrate how this is done via a few examples.

Example 5.1. To start with, we limit ourselves to two gaussian functions $Z_1(x)$ and $Z_2(x)$, to simplify the graphical display. Figure 5.4 shows the 2D gaussian space and the position of two simulated points in this space. As the partition chosen here contains only two subsets, we will have two lithofacies (white and grey). At the first point x_1 the simulated values are $Z_1(x_1)$ and $Z_2(x_1)$, which we will shorten to z_{1-1} and z_{2-1}. This pair of values belongs to the subset labelled F_1. This means that the white facies will be attributed to x_1. At point x_2 the simulated values are z_{1-2} and z_{2-2}. This pair of values belongs to the subset labelled F_2, which means that the

Table 5.1 Proportions of each facies shown in Fig. 4.1

Level (numbered from bottom to top)	Shale (F1) (%)	Shaly sandstones (F2) (%)	Sandstones (F3) (%)
5	100	0	0
4	80	20	0
3	20	80	0
2	0	60	40
1	0	40	60

Table 5.2 Thresholds corresponding to proportions shown in Table 5.1

Level	t_0	t_1	t_2	t_3
5	$-\infty$	∞	∞	∞
4	$-\infty$	0.84	∞	∞
3	$-\infty$	-0.84	∞	∞
2	$-\infty$	$-\infty$	0.25	∞
1	$-\infty$	$-\infty$	-0.25	∞

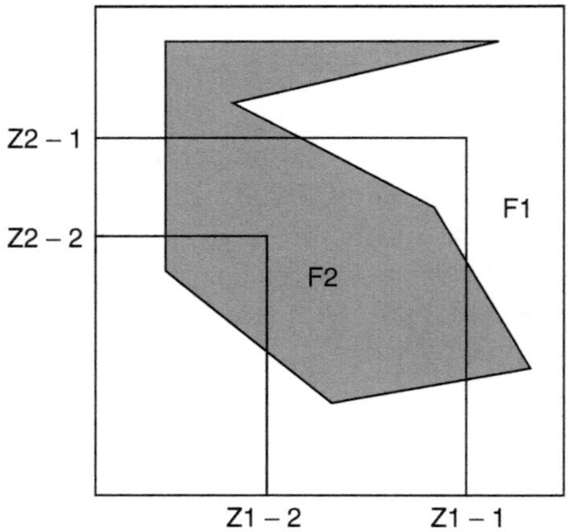

Fig. 5.4 Rocktype rule for two facies

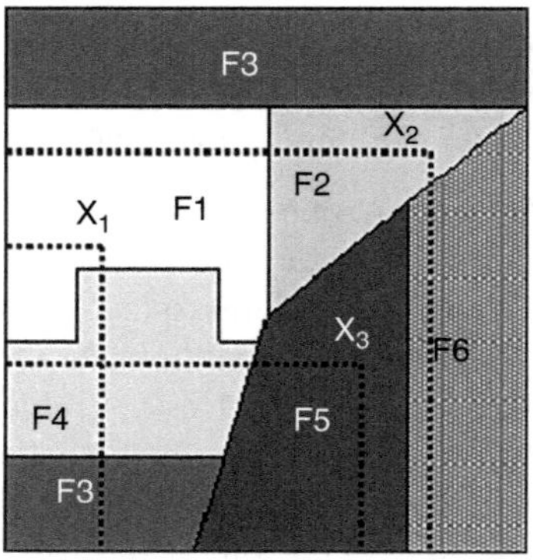

Fig. 5.5 Rocktype rule with six facies. The three points lie in facies N° 1, 2 and 5 respectively

grey facies will be attributed to x_2. (N.B. the gaussian functions take all values between $-\infty$ and $+\infty$. We represent them in a rectangle, to simplify the display.

Example 5.2. In this example, we have also two gaussian functions, but there are six lithofacies. Figure 5.5 shows the 2D gaussian space with its partition, and the position of three simulated points in this space. Here we attribute the facies F_1 to point x_1, F_2 to point x_2 and F_5 to point x_3.

Remark 5.2. 1. In these figures, we see that the limits of the subsets can be rather complicated, so we can no longer speak of thresholds.

2. If we simulate N gaussian functions, they define a theoretical space with N dimensions. At any point x, the values of the simulated gaussian functions, z_1, $z_2, \ldots, z_N$, define the co-ordinates of the simulated value in this new space. The facies are labels attached to parts of this space. If we want to be able to give a facies to each simulated point, these parts must define a partition of the gaussian

space. If we have N gaussian functions Z_j, and if we let D_i be the subset of the gaussian space which is labelled as facies F_i, we can write:

$$1_{Fi}(x) = 1 \Leftrightarrow (Z_1(x), Z_2(x), ..., Z_N(x)) \in D_i.$$

3. The gaussian functions need not to be independent. They can be correlated.

Link with the Proportions

As for the truncated gaussian, the proportion of facies F_i at point x is just the probability of having this facies F_i at that point. It can be written as:

$p_{Fi}(x) = P$ (facies at point $x = F_i) = E[1_{F_i}(x)]$.

As we have $1_{Fi}(x) = 1 \Leftrightarrow (Z_1(x), Z_2(x), ..., Z_N(x)) \in D_i$ we also have:

$$p_{Fi}(x) = P\{[Z_1(x), \ldots, Z_N(x)] \in D_i\} = \int_{D_i} g_\Sigma(z_1, \ldots, z_N)\, dz_1 \ldots dz_N,$$

where $g_\Sigma(z_1, z_2, \ldots, z_N)$ is the N-variate gaussian density function with mean 0 and variance 1, and Σ is its correlation matrix. Computing p_{Fi} is quite easy when we know Σ and D_i, but determining Σ and D_i even when we know all the p_{Fi} is impossible in the general case because there is infinity of solutions. To solve this problem, we have to impose some constraints on the parameters.

Parameter Simplification: Use of Thresholds

To overcome this problem, we choose to partition the gaussian space into rectangles (if we have two gaussian functions) or in rectangular parallelepipeds (with more than two gaussian functions). Figure 5.6 shows an example of one such partition with two gaussian functions. These rectangular boxes are defined by their projections

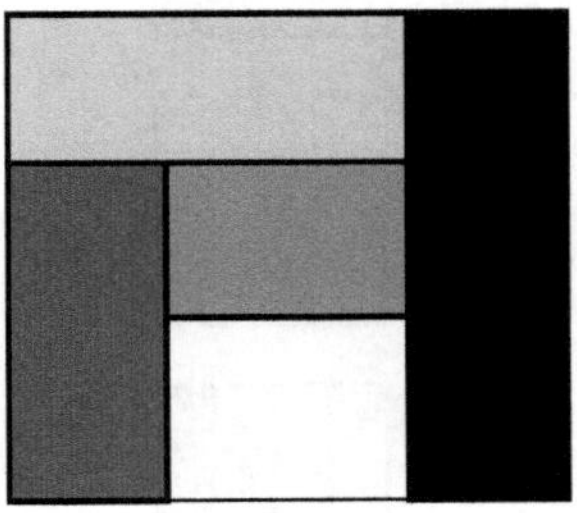

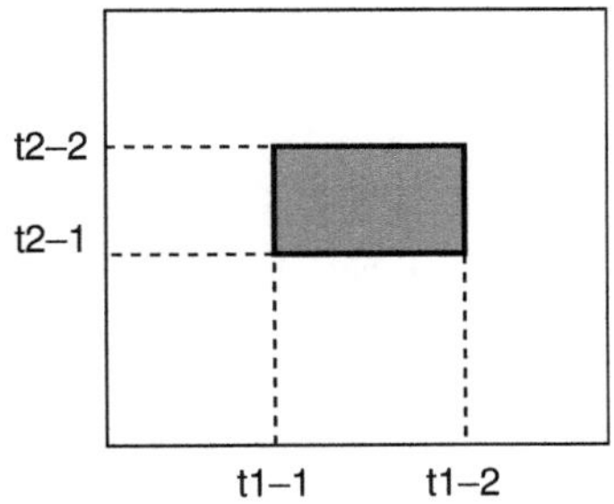

Fig. 5.6 Partition in rectangles (*left*) and thresholds (*right*)

on the gaussian coordinate axes. The end points of these projections can be considered as thresholds. Each rectangular box is defined by 2^N thresholds (some of them can be infinite). When we know $2^N - 1$ of them, it is possible to invert the equation

$$p_{Fi}(x) = \int_{D_i} g(z_1, z_2, \ldots, z_N)\; dz_1\; dz_2 \ldots dz_N$$

numerically to find the last unknown threshold. If we have N gaussian functions and M facies, we have $M \times 2^N$ thresholds. As we do not allow "empty space", our rectangular boxes constitute a partition of this space. This considerably decreases the number of independent thresholds. For example, if we include the infinite thresholds, we have:

- $M + 1$ thresholds with 1 gaussian function (i.e., truncated gaussian method)
- $M + 3$ thresholds with 2 gaussian functions
- $M + 5$ thresholds with 3 gaussian functions
- $M + 2N-1$ thresholds with N gaussian functions

which means $M-1$ finite independent thresholds.

Choice of the Partition

Even after having decided to use a rectangular partition, there are many possible layouts. For example with two gaussian functions and four lithofacies, we have ten possibilities, plus one which is equivalent to the case with only one gaussian function. Figure 5.7 shows these partitions. With 6 facies, we have approximately 140 possibilities. This shows that we will not be able to test all the possible

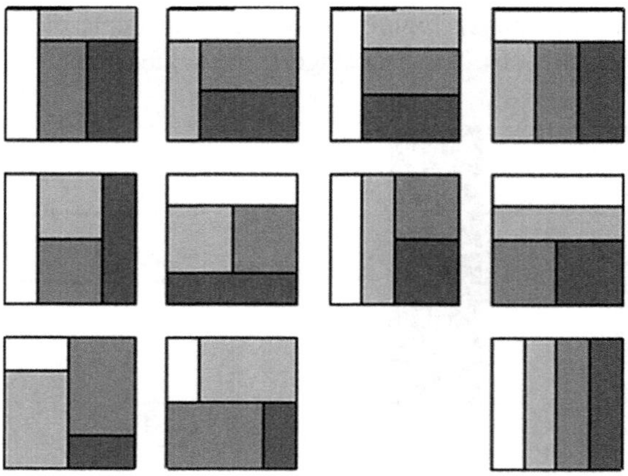

Fig. 5.7 Rectangle rocktype rules for four facies

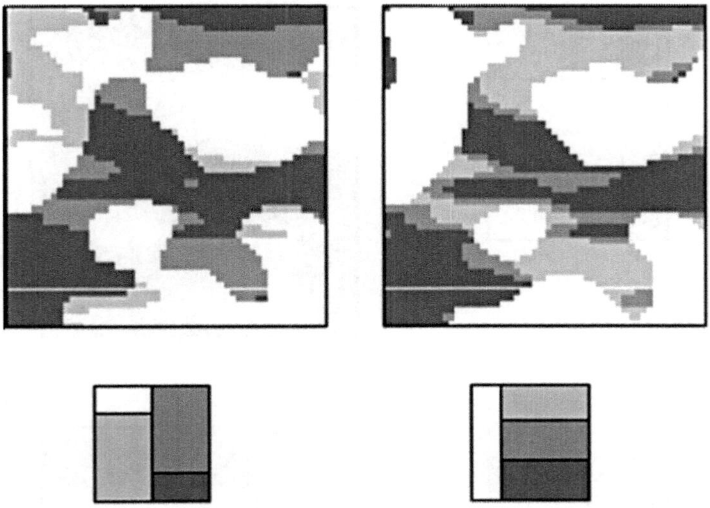

Fig. 5.8 Examples of simulation for two of the rocktype rules shown in Fig. 5.7

partitions. We have to make a choice, and then find a criterion to help us do this. Figure 5.8 shows examples of simulations corresponding to two cases from Figure 5.7.

We can see that the facies that touch each other in the rocktype rule are also in contact in the simulation. This is a general rule when the proportions are constant, and when the simulated field is large enough to be statistically representative.

This rule is usually sufficient to choose the partition if there are only a few facies, but when the number increases we can find several partitions which give the same contact possibilities (Fig. 5.9). In that case, the shape of these contacts can also help us. We have to decide which gaussian function will guide the shape of which contact.

When the proportions vary, some facies can disappear in some parts of the simulated field. In that case, the rocktype rule also varies, and the forbidden contact will locally disappear.

Choice of the Correlation Matrix

With two gaussian functions, the correlation matrix which can be written as

$$\Sigma = \begin{pmatrix} 1 & \rho \\ \rho & 1 \end{pmatrix}$$

is consistent provided that $-1 < \rho < 1$. The values $\rho = \pm 1$ are also authorised, but in that case it is better to work with only one gaussian function. The covariance matrix must always be positive definite. With more than two gaussians, it would be

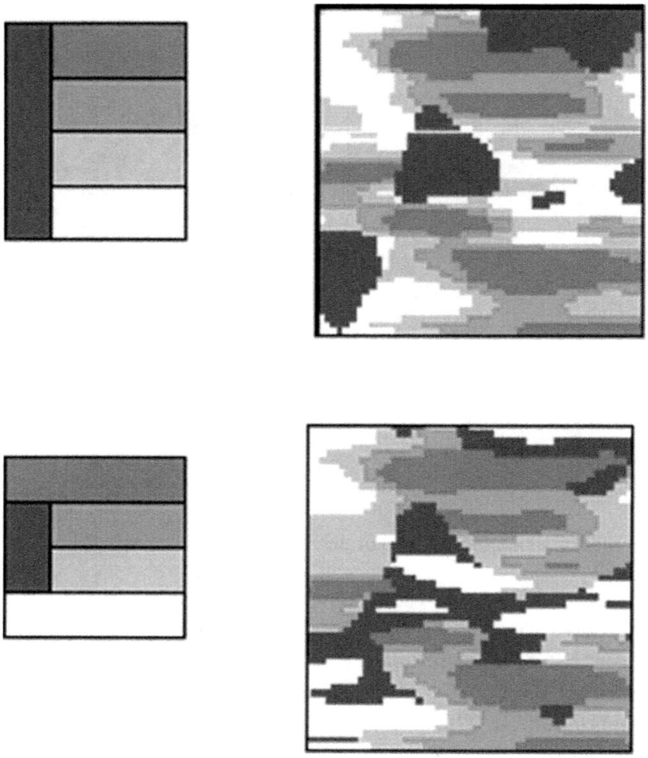

Fig. 5.9 Examples of simulations with the same contacts, but with different rocktype rules

virtually impossible to get a positive definite matrix just by picking values for the terms. It is better to give a consistent model of relationships between the gaussians, and then deduce a suitable covariance matrix from this. We will not go into more detail here.

With two gaussian functions, once the rocktype rule is chosen, we still need to know the $M-1$ thresholds and the correlation, ρ, which gives M unknowns. So we need M equations linking the unknowns to the proportion values. We have M equations but they are not independent because the partition automatically ensures that the proportions sum to 1. As a consequence, we have more variables than independent equations. The number of solutions is infinite. Worse, these solutions are not equivalent: they give rise to simulations that are quite different.

For example Fig. 5.10 shows two simulations obtained with the same partition but different correlations, 0 and 0.6. The shape of the black facies is exactly the same in both cases because the first gaussian function is the same, but those of the three other facies change. This is easiest to see on the white facies. In the right hand simulation where the correlation is 0.6, the white tends to wrap around the black, especially in the bottom left corner. In Chap. 8, we present a case study where correlation will be used to drape one facies over algal bioherms.

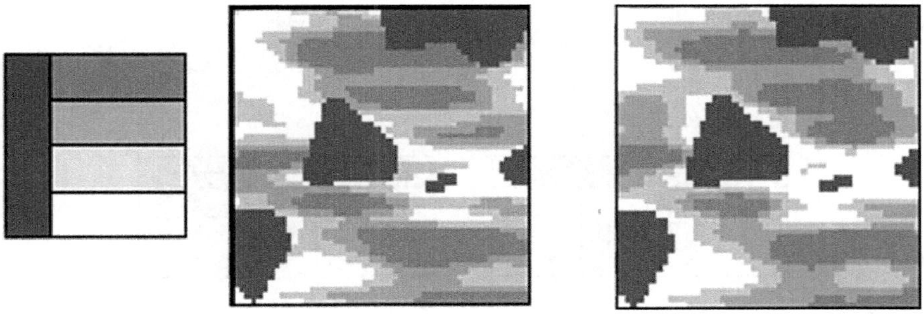

Fig. 5.10 Simulations with the same lithotype rule and different correlation coefficients, $\rho = 0$ and $\rho = 0.6$

The example given above shows that increasing the correlation coefficient introduces a border effect, which looks like the ordering effect we have with only one gaussian function. This is the reason why we suggest choosing the value of the correlation coefficient arbitrarily, depending on whether we want to have a strong border effect or not. Note that the correlation coefficient is a property of the gaussian functions: it will remain constant over the whole domain.

Calculating the Thresholds

When we have chosen the correlation coefficient, we have M−1 independent equations of the form:

$$p_{Fi}(x) = \int_{D_i} g(z_1, z_2, \ldots, z_N)\, dz_1\, dz_2 \ldots dz_N,$$

where the D_i are rectangles $[t_{i-1}, t_i] \times [s_{i-1}, s_i]$.

We have already seen that there are M−1 independent thresholds. This equation system cannot be solved analytically, but iterative methods can give us the solution. The trial and error method, testing successively all the thresholds, would be too slow, because it does not automatically ensure the consistency of the partition in rectangles.

In general, it is better to perform a global optimisation, minimising for example the global square error on the proportions. But in many cases, it is possible to group the facies and work successively on one gaussian function, then the other one (we may have to iterate this procedure). Figure 5.11 shows an example of this. Here the trial and error method is very quick, because the partition is automatically consistent. The partition we want to obtain is shown on the top line. The second line shows the order in which the thresholds are evaluated, starting with the top block.

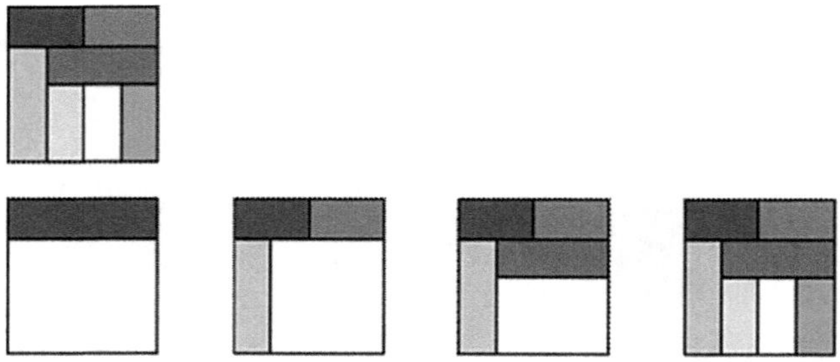

Fig. 5.11 Successive groupings to obtain the thresholds

Generalisation to Non-stationary Case

As for the truncated gaussian, non-stationarity is obtained via varying proportions. This results in varying thresholds for both gaussian functions. The rocktype rule must be given including all the facies, even if some of them can disappear locally.

When Simulations Show "Prohibited" Contacts

Sometimes simulations show contacts which do not exist in the rock type rule diagram. In the stationary case, there are two possible reasons for this: one (or more) of the gaussian functions is discontinuous, or the discretisation is too coarse to show the continuity. This problem occurs much more often in the non-stationary case. We will demonstrate this in the most common case of vertical non stationarity for the truncated gaussian, but the same applies for horizontal non stationarity and in the plurigaussian case.

When one facies disappears, the two facies which should be separated by it come into direct contact. This is obvious when the local rock type rule is shown, but can be forgotten when only the global rocktype rule is given. One consequence is that the relative position of two facies which never appear at the same level is of no importance (see Fig. 5.12). What must be taken into account is their relationship with the other facies.

Two facies which are not in contact in the rock type rule can sometimes touch if the proportions (and hence the thresholds) vary sharply between consecutive levels. This is easy to see on the example below, where there are three facies: shale (F1), shaly sandstone (F2) and sandstone (F3). On level 1, the proportions are respectively 60%, 20% and 20%, which gives the thresholds $t_1 = 0.25$ and $t_2 = 0.84$. On level 2, they are 20%, 20% and 60%, and so the thresholds are $t_1 = -0.84$, $t_2 = -0.25$. The gaussian values between -0.25 and $+0.25$ represent 20% of the histogram. When the value simulated on level 1 is between -0.25 and 0.25 (which corresponds to the facies shale), there is a high probability that on level 2 the

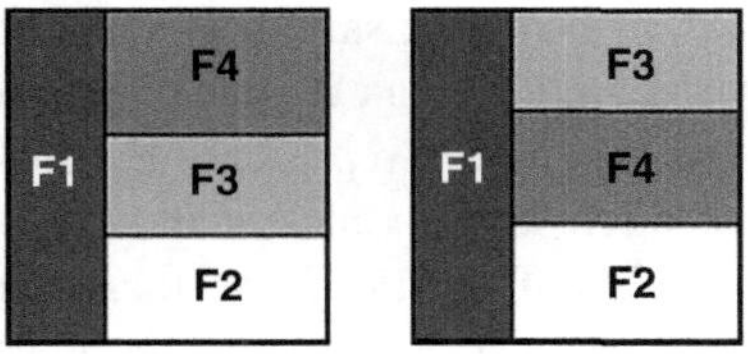

Fig. 5.12 Equivalent rocktype rules if F3 and F4 do not appear on the same level

Table 5.3 Schematic representation of the rocktype rules on two consecutive levels, respecting the ordering in the threshold values

Level 2	Shale	Shaly sandstones		Sandstones	
Level 1		Shale	Shaly sandstones		Sandstones

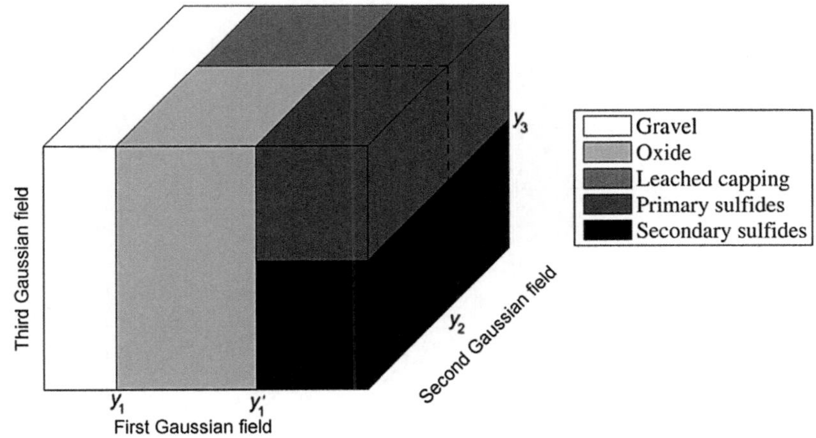

Fig. 5.13 3D rock-type rule used by Emery (2007a, b) for modelling five mineralogical domains in a Chilean porphyry copper deposit: alluvial gravel, oxides, leached capping material, primary and secondary sulphides. Reproduced with permission

simulated gaussian value is located within the same interval. (This of course depends on the variogram model). If that happens, the sandstone will be directly in contact with the shale vertically below it (Table 5.3).

Higher Dimensional Rock-Type Rules

The rock-type rules presented up to this point have all been two dimensional, even though higher dimensional rules are possible as Xu et al. (2006) showed. The reasons for presenting only 2D rules are that they are easier to interpret and to present in papers and reports and because we have not needed higher dimensional rules in any of our case-studies. Having said that, in a study of a Radomiro Tomic porphyry copper deposit north of Calama City in Chile, Emery (2007b) encountered a case where a 3D rule was required to represent the connections between the five mineralogical domains of interest: alluvial gravels, leached capping, oxides, primary and secondary sulphides. The gravel are located near the surface and in contact with the oxides and the leached capping but never the sulphides, but the other four domains are in contact with each other. A convenient way of representing this situation is by using a 3D rule as shown in Fig. 5.13. Introducing a third gaussian allows more flexibility but complicates the variogram analysis.

Chapter 6
Variograms and Structural Analysis

This chapter describes how to calculate experimental variograms for the facies indicators and how to fit models to them. The relationship between the facies indicators and the underlying gaussian values has been given in Chap. 2.

In this chapter we will give the variogram equations in the case when we only have constraints on facies but no other categorical variable or continuous function in the data set.

The theoretical relation linking the variograms of the underlying gaussians and those of the indicators is described. From this link we use a specific fitting method: rather than invert it, we use an indirect iterative procedure to fit a variogram model. But first, we show how to calculate the experimental variograms for the facies indicators.

Experimental Variograms and Cross-Variograms for Facies

In Chap. 2, the definition of the centred covariance of the indicators and its relationship with the gaussian has been given. Here we consider the centred variogram of the indicator of the facies F which is the tool used in most geostatistical studies:

$$\begin{aligned}\gamma_F(x, x+h) &= \frac{1}{2}\mathrm{Var}\left[1_F(x) - 1_F(x+h)\right] \\ &= \frac{1}{2}\left\{E\left(\left[1_F(x) - 1_F(x+h)\right]^2\right) - \left(E\left[1_F(x) - 1_F(x+h)\right]\right)^2\right\}\end{aligned}$$

When the facies indicators are second order stationary or intrinsic with constant mean, the second term is zero and the formula simplifies to:

$$\gamma_F(x,\ x+h) = \gamma_F(h) = \frac{1}{2}E\left(\left[1_F(x) - 1_F(x+h)\right]^2\right)$$

M. Armstrong et al., *Plurigaussian Simulations in Geosciences*,
DOI 10.1007/978-3-642-19607-2_6,

If the mean is not constant, it is still possible to define the non-centred variogram

$$\tilde{\gamma}_F(x, x+h) = \frac{1}{2}E\Big([1_F(x) - 1_F(x+h)]^2\Big)$$

Centred variogram and non-centered variograms are equal when the facies indicators are second order stationary or intrinsic with constant mean. In the non-stationary case, the mean $E[1_F(x)]$ cannot be computed experimentally. In this non-stationary case we have no alternative to the experimental evaluation of the non-centred variogram. As a consequence, we will compute the experimental non centred variogram in all cases.

Hence the formula for computing the experimental variogram for facies F is:

$$\gamma_F^*(x, x+h) = \frac{1}{2N}\sum_{|x_\alpha - x_\beta| = h}\left[1_F(x_\alpha) - 1_F(x_\beta)\right]^2$$

where the data points x_α and x_β are separated by the vector h (possibly with a tolerance). The summation is carried out over all the data pairs separated by this distance, and N is the number of pairs. More information will be given later on how to compute the experimental variogram in the non-stationary case.

In exactly the same way, the cross variogram for facies F_i and F_j is defined as:

$$\gamma_{Fi\,Fj}(x, x+h) = \frac{1}{2}E\big([1_{Fi}(x) - 1_{Fi}(x+h)]\left[1_{Fj}(x) - 1_{Fj}(x+h)\right]\big)$$

Again, this corresponds to the centred (usual) cross-variogram in the second order stationary case, and to a non-centred cross-variogram in the non stationary case. Consequently the formula for the corresponding experimental cross-variogram is:

$$\gamma_{Fi\,Fj}^*(x, x+h) = \frac{1}{2N}\sum_{|x_\alpha - x_\beta| = h}\left[1_{Fi}(x_\alpha) - 1_{Fi}(x_\beta)\right]\left[1_{Fj}(x_\alpha) - 1_{Fj}(x_\beta)\right]$$

Linking the Indicator Variograms to the Underlying Variograms

We have seen in Chap. 2 that the non centred covariance can be interpreted in terms of probability: $\mathrm{Cov}(1_{Fi}(x), 1_{Fj}(x+h)) = P(x \in F_i, (x+h) \in F_j)$

Variograms

From this equation, we can deduce that in the stationary case the variogram of the indicator of facies F_i can also be interpreted in terms of probability:

$$\gamma_{F_i}(x, x+h) = \frac{1}{2}\{P[x \in F_i] + P[x+h \in F_i]\} - P[x \in F_i, x+h \in F_i]$$

If we consider the set of underlying gaussian functions Z(x) and the constraint $\mathcal{C}_i(x)$ which describes the link between Z(x) and the indicator $1_{F_i}(x)$, this equation gives:

$$\gamma_{F_i}(x, x+h) = \frac{1}{2}\{P[Z(x) \in \mathcal{C}_i(x)] + P[Z(x+h) \in \mathcal{C}_i(x+h)]\} - P[Z(x) \in \mathcal{C}_i(x), Z(x+h) \in \mathcal{C}_i(x+h)]$$

Then

$$\gamma_{F_i}(x, x+h) = \frac{1}{2}\{P_{F_i}(x) + P_{F_i}(x+h)\} - \int\limits_{\mathcal{C}_i(x+h)} \int\limits_{\mathcal{C}_i(x)} g_{z(x),z(x+h)}(u, v)du\,dv$$

Cross-Variograms

In the same way as in Chap. 2, we showed that the cross-variogram can be written as

$$\gamma_{F_iF_j}(x, x+h) = -\frac{1}{2}\{E[1_{F_i}(x)1_{F_j}(x+h)] + E[1_{F_j}(x)1_{F_i}(x+h)]\}$$

Hence

$$\gamma_{Fi\,Fj}(x, x+h) = -\frac{1}{2}\Big\{P\big[Z(x) \in \mathcal{C}_i(x),\ Z(x+h) \in \mathcal{C}_j(x+h)\big] + P\big[Z(x) \in \mathcal{C}_j(x),\ Z(x+h) \in \mathcal{C}_i(x+h)\big]\Big\}$$

and

$$\gamma_{Fi\,Fj}(x, x+h) = -\frac{1}{2}\left\{\int\limits_{\mathcal{C}_j(x+h)} \int\limits_{\mathcal{C}_i(x)} g_{z(x),z(x+h)}(u, v)du\,dv + \int\limits_{\mathcal{C}_i(x+h)} \int\limits_{\mathcal{C}_j(x)} g_{z(x),z(x+h)}(u, v)du\,dv\right\}$$

In the next paragraphs, we will explain how to use these general equations in the case of the truncated gaussian method (the vector Z(x) contains only a single gaussian function) and of the classical plurigaussian method (the vector Z(x) contains two or more gaussian functions).

Truncated Gaussian Method

In the truncated gaussian method, $Z(x) \in \mathcal{C}_i(x)$ can also be written $t_{i-1} \leq Z(x) < t_i(x)$ where t_{i-1} and t_i are the thresholds for facies F_i.

In this case,

$$\gamma_{F_i}(x, x+h) = \frac{1}{2}\{P_{F_i}(x) + P_{F_i}(x+h)\} - P[t_{i-1}(x) \leq Z(x) < t_i(x), t_{i-1}(x+h) \leq Z(x+h) < t_i(x+h)]$$

If the correlation function of the gaussian function Z is ρ(h), the indicator variogram is given by:

$$\gamma_{F_i}(x, x+h) = \frac{1}{2}\left\{P_{F_i}(x) + P_{F_i}(x+h)\right\} - \int_{t_{i-1}(x+h)}^{t_i(x+h)} \int_{t_{i-1}(x)}^{t_i(x)} g_{\rho(h)}(u, v))du\,dv$$

$$= \frac{1}{2}\{P_{F_i}(x) + P_{F_i}(x+h)\} - \frac{1}{2\pi\sqrt{1-\rho(h)^2}} \int_{t_{i-1}(x+h)}^{t_i(x+h)} \int_{t_{i-1}(x)}^{t_i(x)} \exp\left(-\frac{u^2 + v^2 - 2\rho(h)uv}{2(1-\rho(h)^2)}\right) du\,dv,$$

where $p_{F_i}(x)$ is the proportion of facies F_i at point x and $g_{\rho(h)}$ is the bigaussian density function. The correlation between the two gaussian variables u and v is ρ(h). Once we have chosen the variogram for the gaussian function, we know ρ(h) and can compute the variogram for each facies using the previous equation.

Cross-Variograms

As the cross-variogram can be written as

$$\gamma_{F_i F_j}(x, x+h) = -\frac{1}{2}\{E[1_{F_i}(x)1_{F_j}(x+h)] + E[1_{F_j}(x)1_{F_i}(x+h)]\}$$

this becomes:

$$\gamma_{F_i F_j}(x, x+h) = -\int_{t_{j-1}(y)}^{t_j(y)} \int_{t_{i-1}(x)}^{t_i(x)} g_{\rho(h)}(u, v)du\,dv$$

Truncated Plurigaussian Method

Variograms

We will give the equations linking the indicator variogram models to those of the gaussians for the case with only two gaussian functions. They can easily be extended to the case where we have more than two gaussians.

In this case, as Z(x) is a bivariate gaussian function, $\mathcal{C}_i(x) \in \mathbb{R}^2$. The variogram can be written:

$$\gamma_{F_i}(x, x+h) = \frac{1}{2}\{P[Z(x) \in \mathcal{C}_i(x)] + P[Z(x+h) \in \mathcal{C}_i(x+h)]\} - P[Z(x) \in \mathcal{C}_i(x),\ Z(x+h) \in \mathcal{C}_i(x+h)]$$

$$\gamma[1_{Fi}(x), 1_{Fi}(x+h)] = \frac{1}{2}\{P_{F_i}(x) + P_{F_i}(x+h)\} - \iint\limits_{\mathcal{C}_i(x+h)} \iint\limits_{\mathcal{C}_j(x)} g_\Sigma(u_1, u_2, v_1, v_2)\, du_1\, du_2\, dv_1\, dv_2$$

If we let Z_1 and Z_2 be the two gaussian functions, components of the bivariate gaussian Z, Σ is the covariance matrix for the four gaussian variables $Z_1(x)$, $Z_2(x)$, $Z_1(x + h)$ and $Z_2(x + h)$ and g_Σ is the quadrivariate gaussian density.

Let ρ be the correlation coefficient between Z_1 and Z_2. Similarly let $\rho_{Z1}(h)$ and $\rho_{Z2}(h)$ be the covariance functions of the two gaussians Z_1 and Z_2. With this information we can compute 12 of the 16 terms in the covariance matrix. The remaining four terms depend on the cross-covariance function between Z_1 and Z_2, $\rho_{Z1Z2}(h)$, which is given by the coregionalization model of Z.

When h represents the vector y−x, the matrix Σ is:

$$\Sigma = \begin{pmatrix} 1 & \rho & \rho_{Z1}(h) & \rho_{Z1\,Z2}(h) \\ \rho & 1 & \rho_{Z1\,Z2}(-h) & \rho_{Z2}(h) \\ \rho_{Z1}(h) & \rho_{Z1\,Z2}(-h) & 1 & \rho \\ \rho_{Z1\,Z2}(h) & \rho_{Z2}(h) & \rho & 1 \end{pmatrix}$$

A wide choice of coregionalization models is available. For example, we can use the standard linear coregionalization model where Z_1 and Z_2 are linear combinations of independent factors. If $Y_1(x)$ and $Y_2(x)$ are independent N(0,1) gaussian functions having the covariances $\rho_{Y1}(h)$ and $\rho_{Y2}(h)$, we can define $Z_1(x)$ and $Z_2(x)$ as

$$Z_1(x) = Y_1(x) \quad \text{and} \quad Z_2(x) = \rho Y_1(x) + \sqrt{1-\rho^2}\, Y_2(x)$$

where

$$\rho_{Z1}(h) = \rho_{Y1}(h) \quad \text{and} \quad \rho_{Z2}(h) = \rho^2\, \rho_{Y1}(h) + (1-\rho^2)\, \rho_{Y2}(h)$$

or

$$\rho_{Y1}(h) = \rho_{Z1}(h) \quad \text{and} \quad \rho_{Y2}(h) = \frac{\rho_{Z2}(h) - \rho^2 \rho_{Z1}(h)}{1-\rho^2}.$$

The cross-covariance is then $\rho_{Z1\,Z2}(h) = \rho\rho_{Z1}(h)$. It is an even function $\rho_{Z1\,Z2}(h) = \rho_{Z1\,Z2}(-h)$, which is not true in general.

The covariance matrix Σ is then:

$$\Sigma = \begin{pmatrix} 1 & \rho & \rho_{Z1}(h) & \rho\rho_{Z1}(h) \\ \rho & 1 & \rho\rho_{Z1}(h) & \rho_{Z2}(h) \\ \rho_{Z1}(h) & \rho\rho_{Z1}(h) & 1 & \rho \\ \rho\rho_{Z1}(h) & \rho_{Z2}(h) & \rho & 1 \end{pmatrix}$$

Alternatively Z_2 could be the derivative of Z_1 (in a given direction) or it could equal the first gaussian shifted by a vector "a". That is,

$$Z_2(x) = \frac{\partial Z_1(x)}{\partial x_i} \quad \text{or} \quad Z_2(x) = Z_1(x - a)$$

Combinations of these are also possible. For example, one model that covers quite a wide range of cases is:

$$\begin{cases} Z_1(x) = Y_1(x) \\ Z_2(x) = \lambda_1 Y_1(x + a) + \lambda_2 Y_2(x) \end{cases}$$

If we want a specific correlation ρ between Z_1 and Z_2, then:

$$\lambda_1 = \frac{\rho}{\rho_{Z1}(a)}, \quad \lambda_2 = \pm\sqrt{1 - \frac{\rho^2}{\rho_{Z1}(a)^2}}$$

In that case, the covariance matrix Σ is:

$$\Sigma = \begin{pmatrix} 1 & \rho & \rho_{Z1}(h) & \dfrac{\rho\rho_{Z1}(h+a)}{\rho_{Z1}(a)} \\ \rho & 1 & \dfrac{\rho\rho_{Z1}(h-a)}{\rho_{Z1}(a)} & \rho_{Z2}(h) \\ \rho_{Z1}(h) & \dfrac{\rho\rho_{Z1}(h-a)}{\rho_{Z1}(a)} & 1 & \rho \\ \dfrac{\rho\rho_{Z1}(h+a)}{\rho_{Z1}(a)} & \rho_{Z2}(h) & \rho & 1 \end{pmatrix}$$

where

$$\rho_{Z2}(h) = \frac{\rho^2}{\rho_{Z1}^2(a)}\,\rho_{Z1}(h) + \left[1 - \frac{\rho^2}{\rho_{Z1}^2(a)}\right]\rho_{Y2}(h)$$

As before, $\rho_{Y2}(h)$ is the covariance function of the second independent gaussian Y_2.

Now we consider the case where the constraints $\mathcal{C}_i(x)$ and $\mathcal{C}_i(x + h)$ are identical, equal to $\mathcal{C}_i$. This corresponds to stationary indicator functions with

$p_{Fi}(x) = p_{Fi}(x + h) = p_{Fi}$). If $\mathcal{C}_i$ are defined as a rectangle limited by the thresholds t_{i1} and t_{i2} for Z_1, and s_{i1} and s_{i2} for Z_2, then the equation becomes:

$$\gamma[1_{Fi}(x), 1_{Fi}(x + h)] = P_{Fi} - \int_{S_{i1}}^{S_{i2}} \int_{t_{i1}}^{t_{i2}} \int_{S_{i1}}^{S_{i2}} \int_{t_{i1}}^{t_{i2}} g_{\Sigma}(u_1, u_2, v_1, v_2)\, du_1\, du_2\, dv_1 dv_2$$

Knowing the proportions, the thresholds and the covariance matrix of Z_1 and Z_2, it is easy to compute (and plot) the indicator variograms.

Cross-Variograms

We saw in Chap. 2 that the cross variogram model can be written as

$$\gamma_{Fi\,Fj}(x, x + h) = -\frac{1}{2}\{E[1_{Fi}(x) 1_{Fj}(x + h)] + E[1_{Fj}(x) 1_{Fi}(x + h)]\}$$

which gives:

$$\gamma[1_{Fi}(x), 1_{Fj}(x + h)] = -\frac{1}{2}\left\{ \iint_{\mathcal{C}_j(x+h)} \iint_{\mathcal{C}_i(x)} g_{\Sigma}(u_1, u_2, v_1, v_2) du_1\, du_2\, dv_1\, dv_2 \right.$$
$$\left. + \iint_{\mathcal{C}_i(x+h)} \iint_{\mathcal{C}_j(x)} g_{\Sigma}(u_1, u_2, v_1, v_2) du_1\, du_2\, dv_1\, dv_2 \right\}$$

In the stationary case where $\mathcal{C}_i(x) = \mathcal{C}_i(y) = \mathcal{C}_i$, this becomes:

$$\gamma[1_{Fi}(x), 1_{Fj}(x + h)] = -\iint_{\mathcal{C}_j} \iint_{\mathcal{C}_i} g_{\Sigma}(u_1, u_2, v_1, v_2) du_1 du_2 dv_1 dv_2$$

Generalisation to the Non-stationary Case

Experimental Variograms

In the non-stationary case, we can only compute the experimental non-centred variogram.

$$\gamma_{exp}(1_{Fi}(x_\alpha), 1_{Fi}(x_\beta)) = \frac{1}{2N} \sum_{|x_\alpha - x_\beta| = h} [1_{Fi}(x_\alpha) - 1_{Fi}(x_\beta)]^2$$

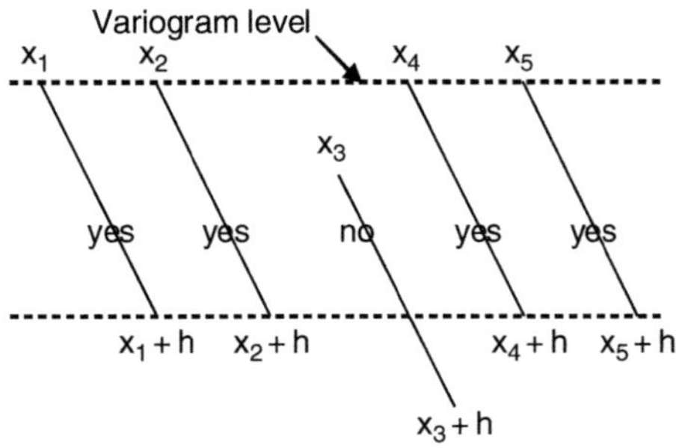

Fig. 6.1 Variogram level

It is no longer equal to the centred variogram. Nevertheless, if the facies can be considered as stationary in the strata plane, the experimental variogram is computed level by level in the plane of the strata (Fig. 6.1). Although the proportions vary along the vertical direction (i.e. vertical after flattening), they can often be considered as constant in the horizontal plane. We call this vertical non-stationarity.

In theory we should fit all these variograms simultaneously (using the same parameters for the gaussian function), but this has two disadvantages:

- In practice, we usually have hundreds of levels, and it takes too much time to check the fit on all these levels;
- If the data come from only a few wells, the statistical fluctuations in the variograms computed level by level are too large, and we cannot fit these variograms at all.

These are the reasons why we average all the variogram levels rather than fitting a model to them all directly. It should be kept in mind that this average is no longer a variogram but a mean variogram. It is still possible to fit it, so we have to find a model for it.

The experimental variogram in the directions of this plane can then be considered as an average of centred variograms (with experimental means which change level by level).

Remarks. In the case of vertical non stationarity, it is necessary to treat the horizontal variogram (average of centred variograms) and the vertical variogram (non centred from start) separately.

Another point to note is that when all the wells cross all the levels vertically, there is no difference between this average level by level and the average in the whole space. Otherwise, these averages are different. We usually compute the average directly throughout the space when the strata cannot be considered as homogeneous.

For the cross variograms, we also compute the non-centred cross variogram

$$\gamma_{\mathrm{exp}}\left(1_{\mathrm{Fi}}(\mathrm{x}), 1_{\mathrm{Fj}}(\mathrm{x}+\mathrm{h})\right) = \frac{1}{2\mathrm{N}} \sum_{\left|\mathrm{x}_\alpha - \mathrm{x}_\beta\right| = \mathrm{h}} \left[1_{\mathrm{Fi}}(\mathrm{x}_\alpha) - 1_{\mathrm{Fi}}(\mathrm{x}_\beta)\right]\left[1_{\mathrm{Fj}}(\mathrm{x}_\alpha) - 1_{\mathrm{Fj}}(\mathrm{x}_\beta)\right]$$

Depending on the type of stationarity (or otherwise), the average is calculated over the whole space, or level by level, as for the simple variograms.

When the cross variograms are computed level by level, we average them again to obtain a mean cross variogram.

Indicator Variogram Model (Truncated Gaussian Model)

We now present the variogram model for the non-stationary case. Earlier in the chapter we saw that for the stationary case, the variogram for the indicator function $1_{Fi}(x)$ is:

$$\gamma_{Fi}(x, x+h) = \frac{1}{2}\{P[t_{i-1} \le Z(x) < t_i] + P[t_{i-1} \le Z(x+h) < t_i]\} \\ - P[t_{i-1} \le Z(x) < t_i, t_{i-1} \le Z(x+h) < t_i]$$

As the thresholds now vary with location, the equivalent equation is:

$$\gamma[1_{Fi}(x), 1_{Fi}(x+h)] = \frac{1}{2}\{P[t_{i-1}(x) \le Z(x) < t_i(x)] + P[t_{i-1}(x+h) \le Z(x+h) < t_i(x+h)] \\ -2P[t_{i-1}(x) \le Z(x) < t_i(x),\ t_{i-1}(x+h) \le Z(x+h) < t_i(x+h)]\}$$

where t_{i-1} and t_i are the thresholds for the facies F_i. This can be written as:

$$\gamma[1_{Fi}(x), 1_{Fi}(x+h)] = \frac{1}{2}\left\{P_{Fi}(x) + P_{Fi}(x+h) - \int_{t_{i-1}(x+h)}^{t_i(x+h)} \int_{t_{i-1}(x)}^{t_i(x)} g_{\rho(h)}(u,v)\,du\,dv\right\}$$

$$= \frac{1}{2}\{P_{Fi}(x) + P_{Fi}(x+h)\} \\ - \frac{1}{2}\frac{1}{\pi\sqrt{1-\rho(h)^2}} \int_{t_{i-1}(x+h)}^{t_i(x+h)} \int_{t_{i-1}(x)}^{t_i(x)} \exp\left(-\frac{u^2+v^2-2\rho(h)uv}{2(1-\rho(h)^2)}\right) du\,dv$$

where $p_{Fi}(x)$ and $p_{Fi}(x+h)$ are the proportions of facies F_i at points x and x + h, and $g_{\rho(h)}$ is the same bigaussian density function as in the stationary case. The cross variogram model is given by:

$$\gamma[1_{Fi}(x), 1_{Fj}(x+h)] = \\ -\frac{1}{2}\left\{\int_{t_{j-1}(x+h)}^{t_j(x+h)} \int_{t_{i-1}(x)}^{t_i(x)} g_{\rho(h)}(u,v)\,du\,dv + \int_{t_{i-1}(x+h)}^{t_i(x+h)} \int_{t_{j-1}(x)}^{t_j(x)} g_{\rho(h)}(u,v)\,du\,dv\right\}$$

Variogram and Cross-Variogram Model (Truncated Plurigaussian Method, two Gaussian Functions)

We have seen that

$$\gamma[1_{Fi}(x),1_{Fi}(x+h)] = \frac{1}{2}\{P_{F_i}(x) + P_{F_i}(x+h)\} - \iint\limits_{C_i(x+h)} \iint\limits_{C_j(x)} g_\Sigma(u_1,u_2,v_1,v_2)du_1\, du_2\, dv_1\, dv_2$$

where Σ is the same covariance matrix as in the stationary case (as the gaussian functions are stationary).

In the case when the constraint $C_i(x)$ is given by a rectangle in $\mathbb{R}^2$, which can be written as:

$$C_i(x) = [t_{i1}(x), t_{i2}(x)] \times [s_{i1}(x), s_{i2}(x)]$$

The variogram equation becomes:

$$\gamma(1_{Fi}(x),1_{Fi}(x+h)) = \frac{1}{2}\left[p_{Fi}(x)+p_{Fi}(x+h) - 2\int_{s_{i1}(x+h)}^{s_{i2}(x+h)}\int_{t_{i1}(x+h)}^{t_{i2}(x+h)}\int_{s_{i1}(x)}^{s_{i2}(x)}\int_{t_{i1}(x)}^{t_{i2}(x)} g_\Sigma(u_1,u_2,v_1,v_2)du_1\, du_2\, dv_1\, dv_2\right]$$

The cross variogram model is given by:

$$\gamma[(1_{Fi}(x),1_{Fj}(x+h))] = \frac{1}{2}\left\{\iint\limits_{C_j(x+h)} \iint\limits_{C_i(x)} g_\Sigma(u_1,u_2,v_1,v_2)du_1\, du_2\, dv_1\, dv_2 + \iint\limits_{C_i(x+h)} \iint\limits_{C_j(x)} g_\Sigma(u_1,u_2,v_1,v_2)du_1\, du_2\, dv_1\, dv_2\right\}$$

Comparing Variogram Models for Indicators and Gaussian Functions

Only the stationary case will be discussed here in detail. In the non stationary case, the proportions vary and this has various effects on the individual variograms. The impact on the mean variogram is much more complex as can be seen from the examples given in the section on "variogram fitting".

Sill of the Indicator Model

In the stationary case, the sill of the indicator variogram for facies F_i is equal to $p_{Fi}(1-p_{Fi})$. As there is no link between the sill of the indicator variograms and that of the gaussians, we can choose any sill for the latter and the standard normal N(0,1) is the simplest choice. In the case of non stationarity, we can no longer speak of a sill. The variogram shape for long distances can be rather complicated; it need not stabilize. This often happens for vertical variograms. This long distance shape is completely controlled by the proportions.

Shape of the Indicator Model

We are now going to see which features of the variogram of the gaussian functions reappear in the indicator variogram. Figure 6.2 shows a gaussian variogram with a range of 50, and the corresponding indicator variogram for the truncated gaussian model. We already know (see Chap. 3) that the indicator variogram cannot have a zero derivative at the origin. We see here that the indicator variogram is linear near the origin. The curvature near the origin which is characteristic of the gaussian variogram has disappeared.

Now look at Fig. 6.3 which shows an exponential variogram with the same practical range of 50, and the corresponding indicator variogram (same proportion and same rock-type rule as in Fig. 6.2). In this case, the overall shape of the indicator variogram for the exponential is closer to that of the exponential than for the gaussian variogram in Fig. 6.2. In general, there is less difference between the shapes of the indicator variograms than between those of the underlying gaussians. So it is not possible to determine the type of model or its parameters from the indicator variogram. In top of this as will be seen in the next section, changes in the proportions have are marked impact on their shape. So the choice between a gaussian variogram and an exponential is made by studying the consequences of this choice on the resulting simulations, especially the continuity and regularity of each facies, rather than the shape of the variogram.

Practical Range of the Indicator Model

The previous figures correspond to the facies F_1 in the rocktype rule given in Fig. 6.4. Figure 6.5 shows the indicator variogram for facies F_2 using the same exponential variogram and the same proportion (33%) as in Fig. 6.3. These figures clearly show that the practical range of the indicator variogram depends on the position of the corresponding facies in the rocktype rule as well as the practical range of the gaussian function variogram.

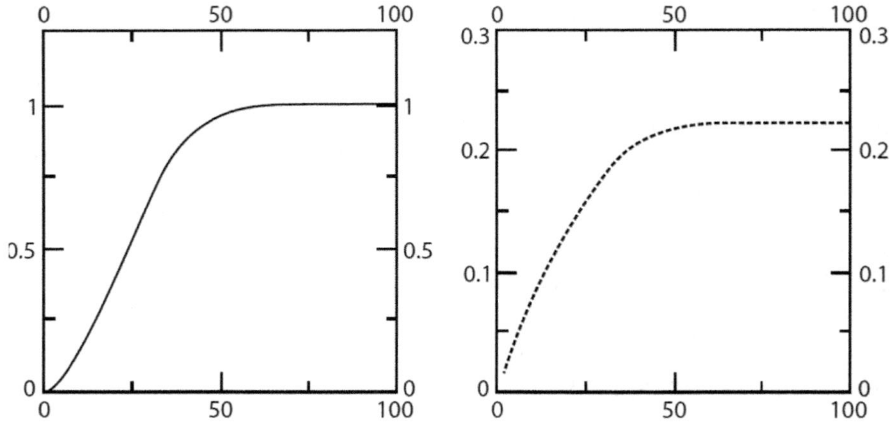

Fig. 6.2 Gaussian variogram (*left*) and the corresponding indicator variogram (*right*)

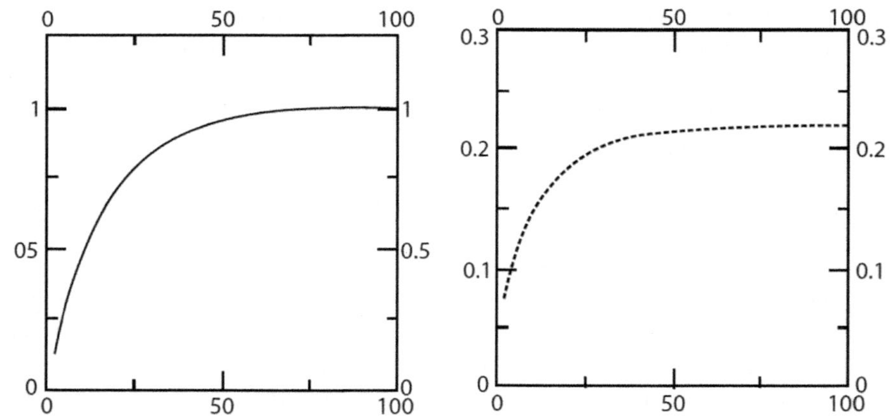

Fig. 6.3 Exponential variogram (*left*) and the corresponding indicator variogram (*right*)

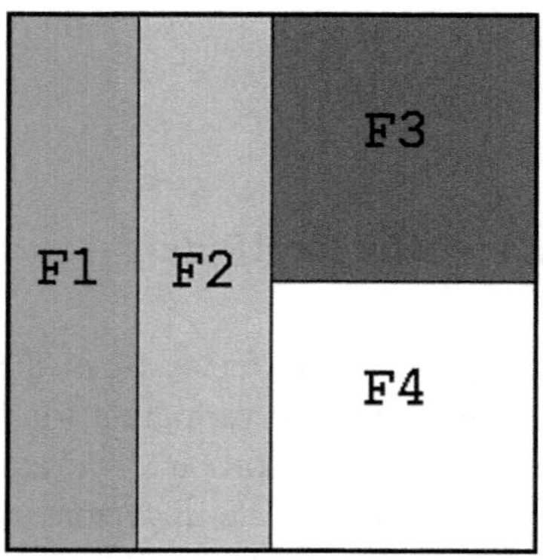

Fig. 6.4 Rock-type rule

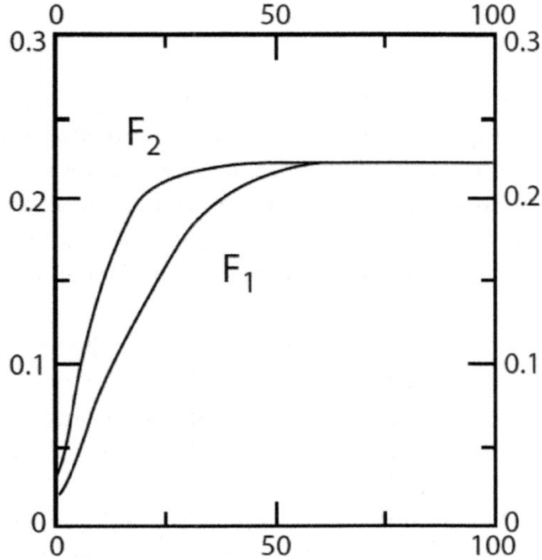

Fig. 6.5 Comparison of the indicator variograms for facies F_1 and F_2

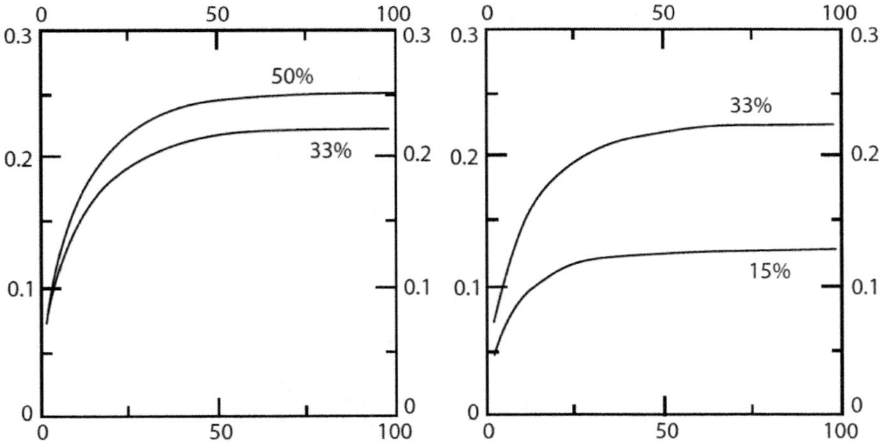

Fig. 6.6 Comparison of the indicator variograms for facies F_2 when the proportion is increased from 33% to 50% (*left*) and decreased from 33% to 15% (*right*)

Figure 6.6 shows what happens to this indicator variogram when the proportion is increased from 33 to 50% or decreased to 15%, respectively. As expected in the first case, the sill increases (a proportion of 50% gives the maximum sill in the stationary case). We see that the practical range has increased too. In the latter case, the practical range decreases as does the sill. This shows that in the same way that we cannot choose the variogram model for the gaussian function just by looking at the shape of the indicator variogram, we cannot deduce its range directly from that of the indicator variogram.

Anisotropies

If the gaussian function model has a geometric anisotropy, the isovalue lines in the variogram map are ellipses. Figure 6.7 shows the isovalue line for the indicator variogram corresponding to 90% of the sill for the rocktype rule depicted in Fig. 6.4. The two gaussian functions have exponential variograms with a geometric anisotropy with ranges of 25 and 100. The long axis of continuity is oriented north-south for the first gaussian function Z_1, and east-west for the other gaussian function Z_2. These are independent.

For facies F_1 and F_2, the isoline curves are still ellipses, as the truncation is carried out on only one gaussian function. The shape is more complicated for facies F_3 and F_4 (as they have the same proportions and are symmetrical in the rocktype rule, their variograms are the same). The result looks like a combination of the two ellipses in perpendicular directions.

Figure 6.8 was obtained in the same way as Fig. 6.7 except that the anisotropy axes have been rotated for Z_2. The long axis is now north-east/south-west. We immediately see that the isolines for facies F_1 and F_2 have not changed, as the truncation for these facies is made only on Z_1 which has not changed. However, the isolines for facies F_3 and F_4 have changed markedly. Here the ellipses with both sets of main anisotropy axes are superposed.

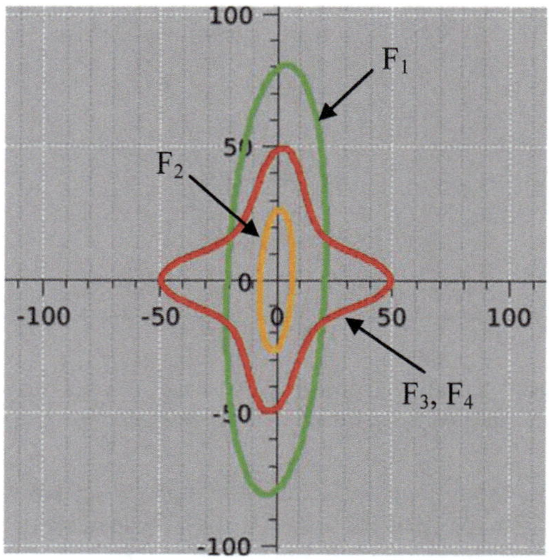

Fig. 6.7 For the facies of Fig. 6.4, isolines representing $\gamma(h) = 90\%$ of the sill. Long anisotropy axes of the gaussian random functions are perpendicular

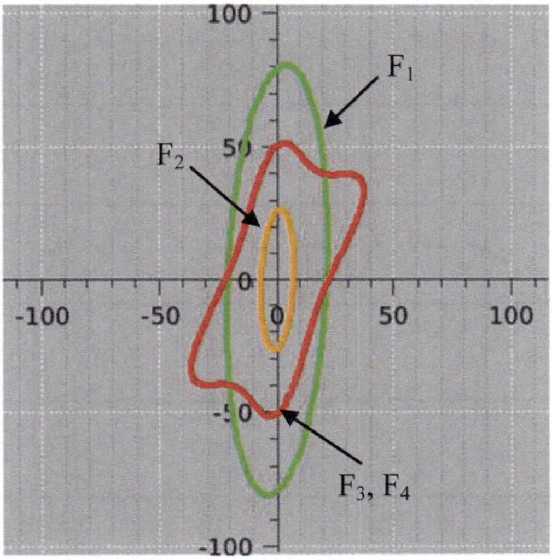

Fig. 6.8 For the facies of Fig. 6.4, isolines representing $\gamma(h) = 90\%$ of the sill. Long anisotropy axes of the gaussian random functions make a 45° angle

Variogram Fitting

As the principle behind variogram fitting is the same in truncated gaussian case and in plurigaussian case, both cases will be presented together.

Stationary Case

We now use the theoretical relation between the variograms of the underlying gaussian variables and the facies indicators to fit models to them. We first choose a model for the gaussian function, to calculate the equivalent indicator model and to plot this. The key point in the variogram fitting is to have good quality experimental variograms in order to compare them with the model. The examples shown in Figs. 6.7 and 6.8 show what is required (at least four directions are required to define the variogram model in a given plane). If the data do not allow us to compute all these experimentally, we need additional information, for example from the geologist.

In the plurigaussian case, if some facies are defined by truncating on one gaussian function only, it is better to fit them first. Once the values of the parameters for this gaussian have been fitted, then the facies for the second gaussian function can be fitted.

Figure 6.9 shows an example of fitting a vertical variogram in the stationary case. Note that this vertical stationarity is not common. In this particular case, experimental vertical proportions show a slight vertical non stationarity, which can be interpreted as statistical fluctuations as there is no clear trend.

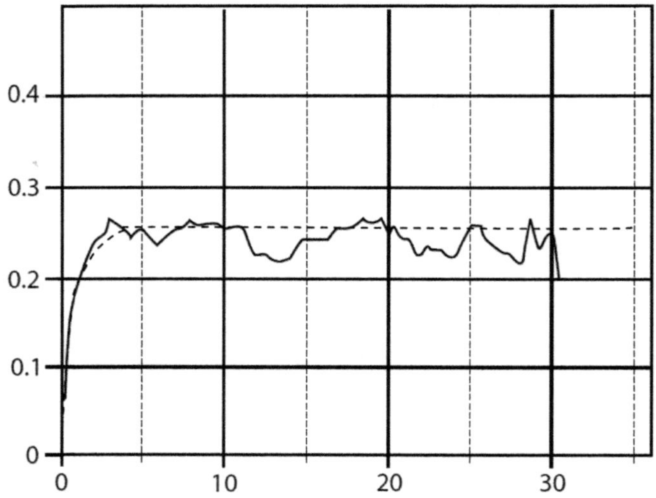

Fig. 6.9 Example of stationary variogram fitting

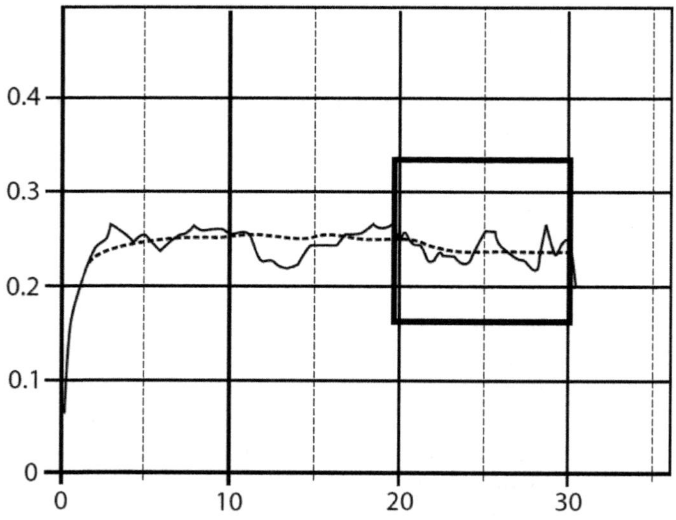

Fig. 6.10 Example of use of the raw varying proportions in a stationary case

Non-stationary Case

The variogram fitting is done the same way as in the stationary case. We have to average the theoretical variograms in the same way as the experimental variograms (level by level or globally). In the non stationary case, the long distance variations are completely controlled by the proportions. Figure 6.10 shows the same

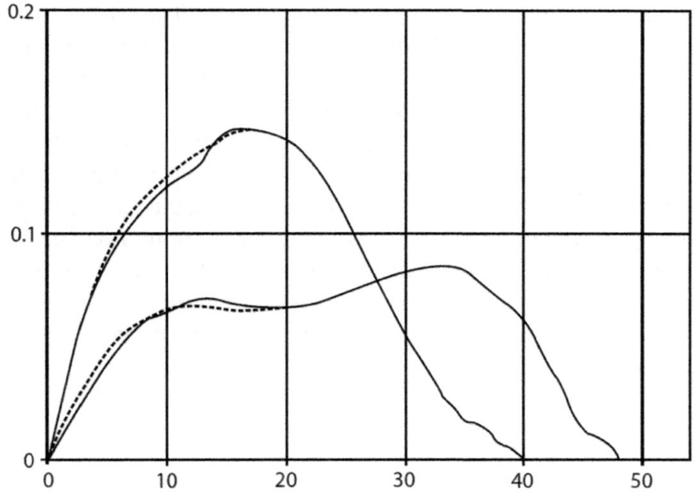

Fig. 6.11 Non stationary variogram fitting using raw proportions

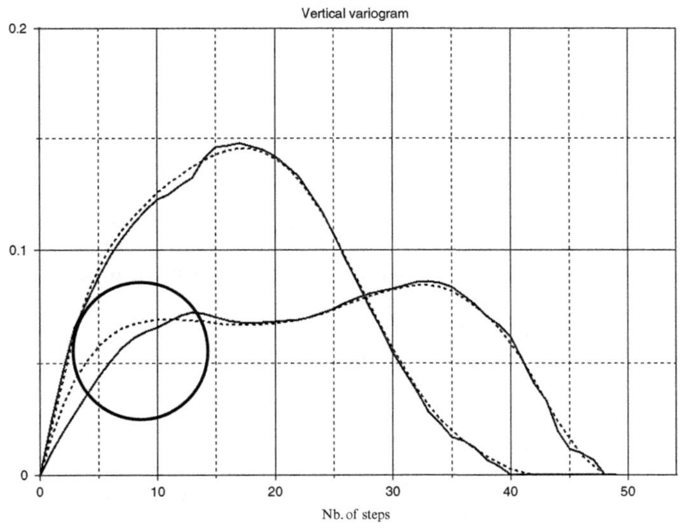

Fig. 6.12 Non stationary variogram computed using smoothed proportions

experimental variogram as in Fig. 6.9, but the slight vertical non stationarity has been taken into account during the fitting process.

Note that the long distances are better fitted here (see black square). Even in the stationary case, it can be easier to fit the variogram with a non stationary approach taking into account the experimental fluctuations of the proportions. The stationary proportions will be used for the simulation, together with the fitted variogram model. If the case is clearly non stationary, we can choose to smooth the proportions for the simulation in order to reduce their experimental fluctuations, but the fitting must be done with the raw proportions. Figure 6.11 shows a variogram fitting in a non stationary case. Its shape is completely irregular, due to the marked variability in the proportions. Figure 6.12 shows the variogram model obtained using the same ranges but computed using smoothed proportions. Smoothing has produced quite marked changes particularly near the origin. In some cases it has been impossible to get a good fit after modifying the proportions. This difficulty can also appear when using proportions which vary in all directions. These proportions usually are the result of estimation and do not come directly from a local average. There can be higher discrepancies between the experimental "variograms" and the model curves. A careful study of the proportions before the variogram fitting stage can reduce these discrepancies.

Transition Probabilities

We saw earlier that transition probabilities are simple to compute when we know the non-centred covariances of the indicator functions:

$$P\left[x + h \in F_j \mid x \in F_i\right] = \frac{C_{ij}(x,\ x + h)}{p_{Fi}(x)}$$

and

$$P\left[x + h \in F_j \mid x \in F_i \text{ and } x + h \notin F_i\right] = \frac{C_{ij}(x,\ x + h)}{p_{Fi}(x) - C_{ii}(x, x + h)}$$

These non centred covariances are easy to compute for the plurigaussian method:

$$C_{ij}(x, x + h) = \iint\limits_{\mathcal{C}_j(x+h)} \iint\limits_{\mathcal{C}_i(x)} g_\Sigma(u_1, u_2, v_1, v_2)\, du_1\, du_2\, dv_1\, dv_2$$

and

$$C_{ii}(x, x + h) = \iint\limits_{\mathcal{C}_i(x+h)} \iint\limits_{\mathcal{C}_i(x)} g_\Sigma(u_1, u_2, v_1, v_2)\, du_1\, du_2\, dv_1\, dv_2$$

This gives the following transition probabilities:

$$P[x + h \in F_j \mid x \in F_i] = \frac{\iint\limits_{\mathcal{C}_j(x+h)} \iint\limits_{\mathcal{C}_i(x)} g_\Sigma(u_1, u_2, v_1, v_2)\, du_1\, du_2\, dv_1\, dv_2}{P_{Fi}(x)}$$

and

$$P[x + h \in F_j \mid x \in F_i \text{ and } x + h \notin F_i] = \frac{\iint\limits_{\mathcal{C}_j(x+h)} \iint\limits_{\mathcal{C}_i(x)} g_\Sigma(u_1, u_2, v_1, v_2)\, du_1\, du_2\, dv_1\, dv_2}{P_{Fi}(x) - \iint\limits_{\mathcal{C}_i(x+h)} \iint\limits_{\mathcal{C}_i(x)} g_\Sigma(u_1, u_2, v_1, v_2)\, du_1\, du_2\, dv_1\, dv_2}$$

We see that these probabilities provide the same information as the non centred covariances. They differ from variograms because they are not symmetrical in general. For example, these probabilities are not symmetrical if the proportions vary in the direction of the vector h, even if the gaussian functions have symmetrical covariances:

$$\iint_{\mathcal{C}_j(x+h)} \iint_{\mathcal{C}_i(x)} g_\Sigma(u_1,u_2,v_1,v_2)\, du_1\, du_2\, dv_1\, dv_2 \neq \iint_{\mathcal{C}_i(x+h)} \iint_{\mathcal{C}_i(x)} g_\Sigma(u_1,u_2,v_1,v_2)\, du_1\, du_2\, dv_1\, dv_2$$

These probabilities can also help to infer the value of the correlation ρ between the gaussian functions, especially the second one:

$$P[x+h \in F_j | x \in F_i \text{ and } x+h \notin F_i] = \frac{C_{ij}(x, x+h)}{P_{Fi}(x) - C_{ii}(x,\ x+h)}$$

Finally, they are easy to compute experimentally in the stationary case provided that there are enough data points to be able to compute reliable statistics. For example, we can only compute the second one when we have enough points within the facies F_i at level x, and enough points which are not in facies F_i at level x + h.

Chapter 7
Gibbs Sampler

Our ultimate objective is to simulate a gaussian random function with a specified covariance structure, given the observed lithotypes (facies) at sample points. As the lithotypes are known at these points, the corresponding gaussian variables must lie in certain intervals or sets but their values are not known. The difficulty is that these random functions conditioned to the constraints are no longer gaussian random functions.

In Chap. 2, the two step procedure used was presented from a theoretical point of view. Here we give a "maths-lite" presentation to illustrate the key concepts. The first section presents several examples to explain why we have recourse to a Gibbs sampler to generate gaussian values at sample points that have the right covariance and belong to the right intervals. Once we have this set of point values, any method for conditionally simulating gaussian random functions can be used; for example, turning bands together with a conditioning kriging, sequential gaussian simulations, LU decomposition, etc. See Chilès and Delfiner (1999), Lantuéjoul (2002a, b) or Deutch and Journel (1992). As these techniques are well known, we will not dwell on them here.

Why We Need a Two Step Simulation Procedure

The aim of this section is to highlight the difficulties of simulating gaussian random functions subject to interval constraints. To do this we consider three simple cases where there are only two points:

- With no constraints
- With interval constraints on one variable
- With interval constraints on both variables

In the first case, the conditional distribution of Z(x) given Z(y) turns out to be a gaussian distribution but this is no longer true in the other two cases. The conditional distributions are merely proportional to gaussians.

M. Armstrong et al., *Plurigaussian Simulations in Geosciences*,
DOI 10.1007/978-3-642-19607-2_7, © Springer-Verlag Berlin Heidelberg 2011

Simulating Z(x) and Z(y) When There Are No Constraints

Consider two gaussian variables Z(x) and Z(y) with a correlation coefficient, ρ. In order to simulate Z(x) given Z(y) we need to know its conditional distribution which can be deduced from the joint distribution of the two variables:

$$g(u,v) = \frac{1}{2\pi\sqrt{1-\rho^2}} \exp\left\{-\frac{(u^2+v^2-2\rho uv)}{2(1-\rho^2)}\right\}$$

where u and v represent z(x) and z(y) respectively. This can be rewritten as

$$g(u,v) = \frac{1}{\sigma\sqrt{2\pi}} \exp\left\{-\frac{(u-\rho v)^2}{2\,\sigma^2}\right\} \frac{1}{\sqrt{2\pi}} \exp\left\{\frac{-v^2}{2}\right\} \tag{7.1}$$

where $\sigma^2 = 1 - \rho^2$. The second term is just the marginal distribution of Z(y). The first is the conditional distribution that we are looking for. It is clearly a gaussian distribution with mean, ρv, and variance σ^2. Equation (7.1) can be written as

$$g(u,v) = g_v(u)g(v) \tag{7.2}$$

This is equivalent to the well-known decomposition:

$$Z(x) = \rho Z(y) + \sigma R(x)$$

where R(x) is a N(0,1) residual that is independent of Z(y). If we estimate Z(x) given z(y), the simple kriging weight equals ρ, and the SK variance is $\sigma^2 = 1 - \rho^2$.

To simulate pairs of values of Z(x) and Z(y), we first draw two independent N(0,1) values for Z(y) and R(x), then we substitute them into the decomposition formula to get Z(x). Alternatively we could say that we draw one realisation v of a N(0,1) variable for Z(y), followed a N(ρv,σ^2) variable for Z(x). In that case, we use the marginal distribution of Z(y) to draw a realisation of it, then the conditional distribution of Z(x) given Z(y) to draw the other value directly. This is only possible because the form of the conditional distribution is so simple.

Simulating Z(x) and Z(y) When Z(y) Belongs to an Interval

In this case Z(y) is known to lie in a specified interval, I, and we want to simulate the pair of variables, Z(x) and Z(y), given that Z(y) lies in that interval. The joint density of the two variables is now

$$h(u,v) = k\, g(u,v) 1_I(v)$$

where $1_I(v)$ is the indicator function for the interval I and k is the normation factor required to ensure that the integral of h(u,v) sums to 1. So the joint density is

$$h(u,v) = \frac{1}{\sigma\sqrt{2\pi}} \exp\left\{\frac{-1}{2}\frac{(u-\rho v)^2}{\sigma^2}\right\} \times \frac{k}{\sqrt{2\pi}} \exp\left\{\frac{-1}{2}\frac{v^2}{2}\right\} 1_I(v)$$

This can be written as

$$h(u,v) = h_V(u)\, h(v)$$

It is clear that h(v) is the marginal distribution of a gaussian variable Z(y) restricted to the interval I. To show that it is the marginal distribution, we just have to prove that

$$\int_{\Re} h_v(u)\, du = 1$$

Integrating h(u, v) with respect to u gives

$$\int_{\Re} h(u,v)\, du = \int_{\Re} \frac{1}{\sigma\sqrt{2\pi}} \exp\left\{-\frac{(u-\rho v)^2}{2\sigma^2}\right\} du \quad \frac{k}{\sqrt{2\pi}} \exp\left\{-\frac{v^2}{2}\right\} 1_I(v)$$

As the first term on the right hand side is just the integral of a $N(\rho v,\sigma^2)$ variable, it equals 1, which gives us the required result. In order to simulate these we use the same interpretation as before: first simulate Z(y) in the interval I then simulate Z(x) given that $Z(y) \in I$. The first simulation is just a truncated gaussian; the second one corresponds to simulating an independent $N(\rho v,\sigma^2)$ variable. That is, we are still using the classical decomposition.

But there is a fundamental change in the marginal distribution of Z(x). To see this, we integrate the joint density with respect to v using (7.2) written as $g(u,v) = g_u(v)g(u)$:

$$\int_{\Re} h(u,v)\, dv = \int_{\Re} k\, g(u,v)\ \ 1_I(v)\, dv$$

$$= \int_{\Re} \frac{1}{\sigma\sqrt{2\pi}}\, 1_I(v)\ \exp\left\{-\frac{(v-\rho u)^2}{2\sigma^2}\right\} dv \times \frac{k}{\sqrt{2\pi}} \exp\left\{-\frac{u^2}{2}\right\}$$

If we let $t = v - \rho u$, then $v \in I \Leftrightarrow t \in I - \rho u$ and the first term on the right becomes

$$\int_{\Re} \frac{1}{\sigma\sqrt{2\pi}}\, 1_{I-\rho u}(t) \exp\left\{-\frac{t^2}{2\sigma^2}\right\} dt = E\left[1_{I-\rho u}\right]$$

As $E[1_{I-\rho u}] = P_\sigma[I - \rho u]$

$$\int_{\Re} h(u, v)dv = k\frac{1}{\sqrt{2\pi}}\exp\left\{\frac{-u^2}{2}\right\} \times P_\sigma(I - \rho u)$$

This shows that the marginal density of Z(x) given that Z(y) $\in$ I, is no longer gaussian. It is proportional to a gaussian density but is multiplied by the probability that the first variable lies in the interval, I − ρu. To illustrate the impact of this change, we have plotted this distribution for two intervals: [−0.5, 0.5] and [2, 3], and for two different correlation factors, −0.5 and 0.8. Figures 7.1 and 7.2 present the resulting curves.

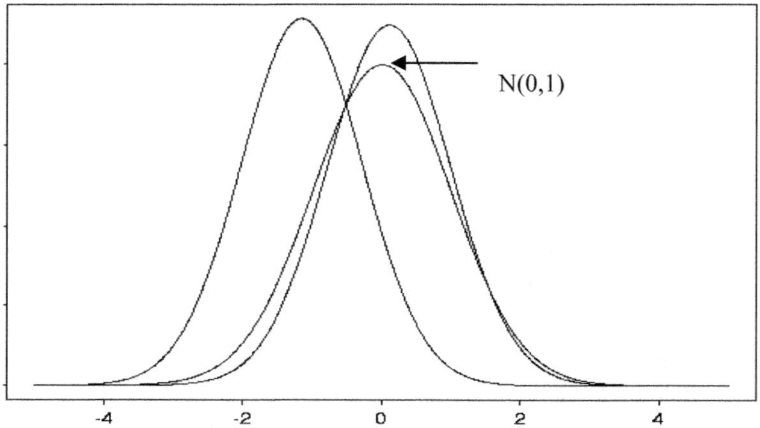

Fig. 7.1 Probability densities functions for the marginal distributions for the two intervals, for the case where $\rho = -0.5$ together with the N(0,1) density for comparison purposes

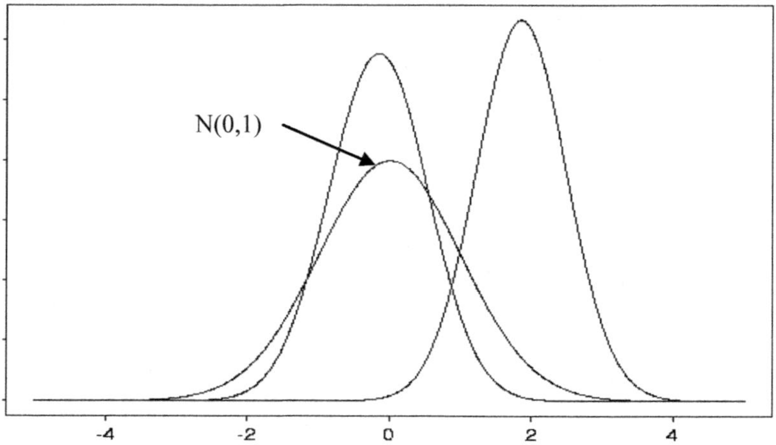

Fig. 7.2 Probability densities functions for the marginal distributions for the two intervals, for the case where $\rho = 0.8$ together with the N(0,1) density for comparison purposes

Simulating Z(x) and Z(y) When Both Belong to Intervals

Now suppose that Y(x) belongs to I_1 and Y(y) belongs to I_2. Their joint density is

$$h(u,v) = kg(u,v)\, 1_{I_1}(u)\, 1_{I_2}(v)$$

where k is the appropriate normation factor.

$$h(u,v) = \frac{k}{\sigma\sqrt{2\pi}} \exp\left\{-\frac{(u-\rho v^2)}{2\sigma^2}\right\} 1_{I_1}(u) \frac{1}{\sqrt{2\pi}} \exp\left\{-\frac{v^2}{2}\right\} 1_{I_2}(v)$$

As expected, the marginal distributions are no longer gaussian or even truncated gaussian. Integrating with respect to u gives

$$\int_{\Re} h(u,v)\, du \propto \frac{1}{\sqrt{2\pi}} P_{\sigma_2}(I_1 - \rho v) \exp\left\{\frac{-v^2}{2}\right\} 1_{I_2}(v) \tag{7.3}$$

Similarly integrating with respect to v gives

$$\int_{\Re} h(u,v)\, dv \propto \frac{1}{\sqrt{2\pi}} P_{\sigma_1}(I_2 - \rho u) \exp\left\{\frac{-u^2}{2}\right\} 1_{I_1}(u) \tag{7.4}$$

Because of the increasing difficult in simulating these distributions directly as the number of interval constraints increases, we rapidly reach the point where direct simulation is no longer practicable and we have to resort to an indirect approach such as the Gibbs sampler.

These examples suggest the idea of using a two step procedure to conditionally simulate gaussian values when some of them are constrained to lie in specified intervals. The two steps are:

1. Generating gaussian values at data points, in the prescribed intervals and with the right covariance structure
2. Using any algorithm for conditionally simulating gaussian random functions given the values generated in step (1)

Direct Simulation Using an Acceptance/Rejection Procedure

Having seen that the first step is to generate a set of gaussian values at sample points, the next question is how this should be done. We might be tempted to try an acceptance/rejection procedure, by generating gaussian values with the right covariance structure and rejecting those lying outside the specified intervals. If only 10 or 20 samples were available this could be done directly using an LU decomposition (i.e. a Cholesky decomposition of the covariance matrix). The LU decomposition could be carried out for much larger matrices (up to about 500 $\times$ 500). The limiting factor in the procedure is the rate of rejection. For example, suppose that ten samples were available and that there were only two facies each present 50% of the time. Then if there was no spatial correlation (pure nugget effect), the probability of getting all ten values in the right intervals would be 1 in 2^{10}; that is, about 1 in 1,000. This procedure becomes prohibitively slow as the number of samples increases. As there are usually hundreds or thousands of data in mining and petroleum applications, another approach is needed. This is why we have to resort to more complicated methods.

Gibbs Sampler

Statisticians routinely use iterative methods based on Markov chain Monte Carlo simulations (MCMC, for short) for sampling complicated distributions and for estimating parameter values. The best known are the Hastings-Metropolis algorithm and the Gibbs sampler. The latter is a particular case of the Hastings-Metropolis method. See Meyn and Tweedie (1993), Cowles and Carlin (1996) and Robert (1996) for information on these methods. Freulon (1992) and Freulon and de Fouquet (1993) adapted the Gibbs sampler to truncated gaussian simulations. To introduce this technique, we present an example to show how this method works and to illustrate the concept of convergence.

Four Sample Example

Suppose that there are only two lithotypes F_1 and F_2, and that they are present in equal proportions. So it is natural to use a zero threshold to separate them. Negative

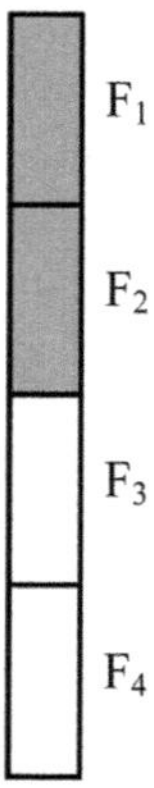

Fig. 7.3 Simplified well or drill hole containing four samples

gaussian values correspond to F_1; positive ones, to F_2. Figure 7.3 shows a simplified well or drill hole containing four samples, with the top two belonging to F_1 and the other two belonging to F_2. The gaussian values assigned to the top samples must be negative, the others have to be positive.

The procedure relies on the standard decomposition of the gaussian random function into its simple kriging estimate and an orthogonal residual.

$$Z(x) = Z_{SK}(x) + \sigma_{SK} R(x)$$

where the index SK denotes the simple kriging estimate or its variance and R(x) is a N(0,1) residual. Note that for gaussian random variables, the SK estimate equals the conditional expectation.

Exponential Variogram

Suppose that the four samples are 1m apart and that the underlying variogram for the gaussians is an exponential with a unit sill and a scale parameter a = 2 m (i.e. a practical range of 6 m). The SK weights for estimating the top point using the other three points as data are:

$$\lambda_2 = 0.61, \quad \lambda_3 = 0, \quad \lambda_4 = 0 \quad \text{and} \quad \sigma^2_{SK} = 0.63 \Rightarrow \sigma_{SK} = 0.79$$

(Points are numbered from the top down). By symmetry the weights are the same but in the reverse order when kriging the fourth point from the other three. The weights for the second point (or similarly the third one) are:

$$\lambda_1 = 0.44, \quad \lambda_3 = 0.44, \quad \lambda_4 = 0 \quad \text{and} \quad \sigma^2_{SK} = 0.46 \Rightarrow \sigma_{SK} = 0.68$$

As the configuration does not change from one iteration to the next the weights remain the same.

Step 1: Initialising the Procedure

The first step consists of choosing gaussian values that belong to the appropriate intervals. Here we select (−1, −1, +1, +1)

Step 2: Iterative Procedure

Simple kriging is applied to the points in turn. For example, the value of the top point is kriged using the other three points as input data. Then we move down to the second point and krige it using the initial values for the points below it and the new updated value for the top point. After completing the second point we move down to the third one which is kriged using the updated values for the points above it and the old value for the point below it. Similarly for the fourth point. When all the points have been updated by kriging, one iteration has been completed.

Point No 1

The kriged estimate for $Z(x_1)$, abbreviated to Z(1), based on the initial values (i.e. −1, +1, +1) for the other three points is:

$$Z_{SK}(1) = -1 \times 0.61 + 1 \times 0 + 1 \times 0 = -0.61$$

And the corresponding residual must satisfy

$$R(1) \leq -(-0.61)/0.79 = 0.77$$

Suppose for argument's sake that we draw a value of 0.52 (from a N(0,1) distribution). Then the updated value would be

$$Z(1) = -0.61 + 0.79 \times 0.52 = -0.20$$

Point No 2

The kriged estimate for Z(2) based on the initial values for Z(3) and Z(4), and the updated value of Z(1) is:

$$Z_{SK}(2) = -0.20 \times 0.44 + 1 \times 0.44 + 1 \times 0 = 0.42$$

The corresponding residual must satisfy

$$R(2) \leq 0.42/0.68 = 0.62$$

If we draw a value of −0.75, then the updated value of Z(2) is

$$Z(2) = 0.42 + 0.68 \times -0.75 = -0.09$$

Point No 3

Following the same procedure, the kriged estimate for Z(3) and the inequality to be satisfied by its residual are

$$Z_{SK}(3) = -0.20 \times 0 - 0.09 \times 0.44 + 1 \times 0.44 = +0.40$$

$$R(3) \geq 0.40/0.68 = 0.59$$

If we draw a value of 0.28, then the updated value of Z(3) is 0.21.

Point No 4

In the same way, the kriged estimate for Z(4) and the inequality to be satisfied by its residual are

$$Z_{SK}(4) = 0 \times (0.20) + 0 \times (-0.09) + 0.61 \times 0.21 = 0.13$$
$$R(4) \geq -0.13/0.79 = -0.16$$

Drawing a value of 0.63 would give an updated value of 0.63 for Z(4).

Results

Table 7.1 summarises the intermediate results during first iteration. This updating procedure is repeated iteratively, in general for several hundred or several thousand iterations. Table 7.2 shows the results of the first five iterations.

Alternative Updating Strategies

In the previous example, individual points were sequentially updated. A variant of this consists of sequentially updating from the top down, then from the bottom up on the next iteration.

Table 7.1 Successive steps in the first iteration of this Gibbs sampler

F1	-1	-0.20	-0.20	-0.20	-0.20
F1	-1	-1	-0.09	-0.09	-0.09
F2	+1	+1	+1	+0.21	+0.21
F2	-1	-1	-1	-1	+0.63

Table 7.2 Results of first five iterations of the Gibbs sampler

Pt N°	Initial	No 1	No2	No3	N° 4	N° 5
1	−1	−0.20	−0.37	−1.34	−0.34	−0.13
2	−1	−0.09	−0.35	−0.24	−0.31	−0.20
3	+1	+0.21	+0.15	+0.05	+0.19	+0.20
4	−1	+0.63	+0.86	+0.10	+0.26	+0.32

Blocking Factor

It is also possible to update blocks of points simultaneously. For example, the four points could be grouped into two blocks each consisting of two points. There are three possible groupings of this type:

- Points 1 & 2 and Points 3 & 4
- Points 1 & 3 and Points 2 & 4
- Points 1 & 4 and Points 2 & 3

In the first case, the values of the top two points are updated using the values of the other two as the conditioning data (i.e. using kriging) and simulated in the right interval with the right correlations, and vice versa for the other pair. We will illustrate this procedure later in the chapter. As was shown in Chap. 2 suitably chosen blocking strategies can significantly improve the speed of convergence.

Experimentally Testing Convergence

Having seen how the procedure works, several questions need to be answered. Firstly, does the algorithm converge? If so, after how many iterations? What factors affect the speed of convergence? How should we choose the initial values?

Burn-in Period

In this section we illustrate the difference between the initial burn-in period and the subsequent stationary part of the Markov chain. To do this we continue the previous example but the variogram is changed to a gaussian model with a practical range of 3. So the correlation between adjoining samples is 0.95. Five hundred iterations of the corresponding Gibbs sampler were run starting from a very extreme set of initial values (+5,+5,−5,−5). This choice lengthens the initial burn-in period, making it visually much more obvious.

Figure 7.4 shows the output for each component as a function of the number of iterations. The values of the first component (top left) decrease steadily from the initial value of $+5$ until they are below 1.0. The curve seems to stabilise after approximately 100 iterations so the burn-in period must be at least this long. Similarly for the third and fourth components. But it appears to be much shorter for the other component (bottom left), about 20 iterations. This shows that the burn-in period need not be the same for all components in a Gibbs sampler. In MCMC theory it is well-known that different states can have different rates of convergence; see Meyn and Tweedie (1993, pp 362–363).

The implications of not necessarily having the same burn-in period for all components are important in practice. When there are only four components it is

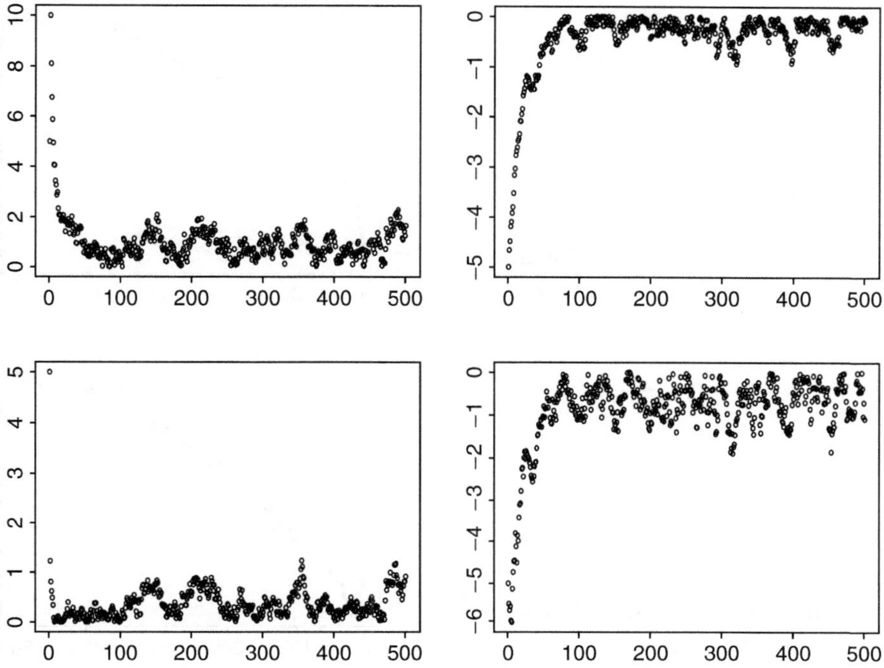

Fig. 7.4 Output of the Gibbs sampler for all 4 components for 500 iterations, starting from initial values of (+5,+5,−5,−5). The second component (*lower left*) seems to stabilise after about 20 iterations components whereas the other three are much slower. They take at least 100 iterations to reach their stationary distribution

possible to check the convergence of all of them but if there were 1,000 samples it would be virtually impossible to inspect the output of the Gibbs sampler for all 1,000 components. We could only check a few of them visually and we might have the bad luck to choose those with shorter burn-ins. Inspecting the results for selected components gives us an idea of the burn-in period but it is not foolproof.

Effect of the Range on the Burn-in Period

Several factors including the range of the variogram and the number of components have a marked effect on the length of the burn-in period. To illustrate the effect of the range, we repeated the previous example using a practical range of 5 instead of 3. This increases the correlation between adjoining samples from 0.95 to 0.98. Figure 7.5 shows the output.

Whereas the burn-in period for the first component was about 100 beforehand, it is now closer to 500. Conversely decreasing the range would decrease the burn-in period. Looking at Fig. 7.5 we also notice how smooth the curves are

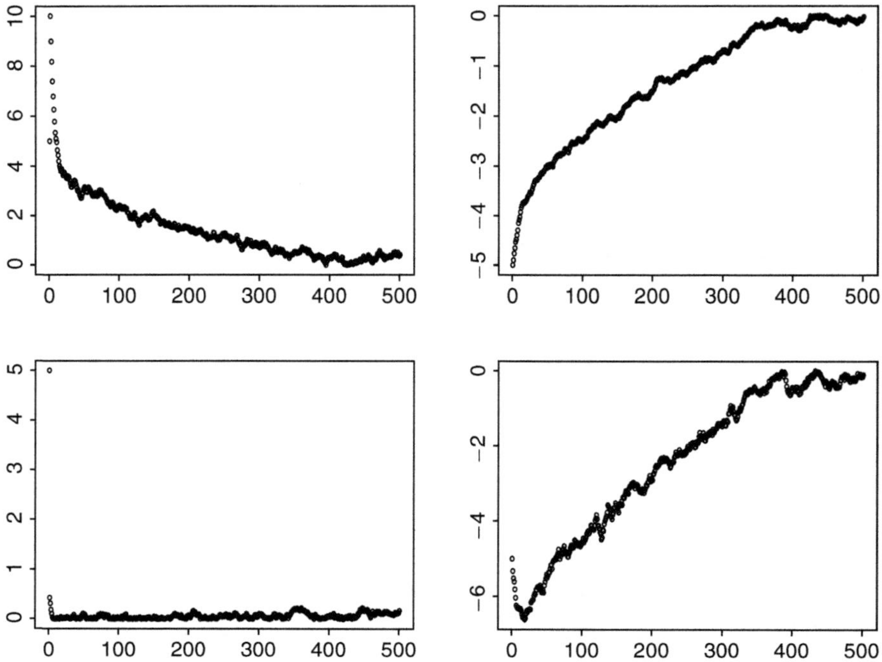

Fig. 7.5 Output of the Gibbs sampler for the first of 4 components for 500 iterations, starting from initial values of (+5,+5,−5,−5). Compared to Fig. 7.4, the correlation between adjoining points has been increased from 0.95 to 0.98. Note the increase in the burn-in period

compared to the corresponding ones in Fig. 7.4. The strong serial correlation between successive values makes it more difficult to determine whether the Markov chain has converged.

These two examples show that it is not simple to judge whether a Gibbs sampler has reached its stationary distribution just by studying the output from a single run (even a very long one). It would be better to run a large number of samplers in parallel and study their output after 1, 5, 10, ..., 50 iterations and so on. Ideally we should compare the experimental distribution of the output with the stationary distribution. How could this be tested experimentally?

A multivariate normal distribution is fully specified when the means and the covariance matrix are known. By extension, a truncated gaussian is fully defined when the truncation thresholds are known together with the means and the covariance matrix for the full distribution. Having said that, it is clear that after truncation the means are not going to be the same as before. See (7.3) and (7.4). Nor are the variances or the correlations. In the case under study, the four components come from a quadrivariate normal distribution with the covariance matrix given in Table 7.3.

Because of the high correlation between adjoining samples, its density is an elongated cigar shape. Figure 7.6 shows a diagram representing the bivariate densities of X_1 and X_2, and X_2 and X_3 respectively. The impact of different types

Table 7.3 Quadrivariate covariance matrix for the 4-sample case

$$\begin{bmatrix} 1 & 0.95 & 0.80 & 0.61 \\ 0.95 & 1 & 0.95 & 0.80 \\ 0.80 & 0.95 & 1 & 0.95 \\ 0.61 & 0.80 & 0.95 & 1 \end{bmatrix}$$

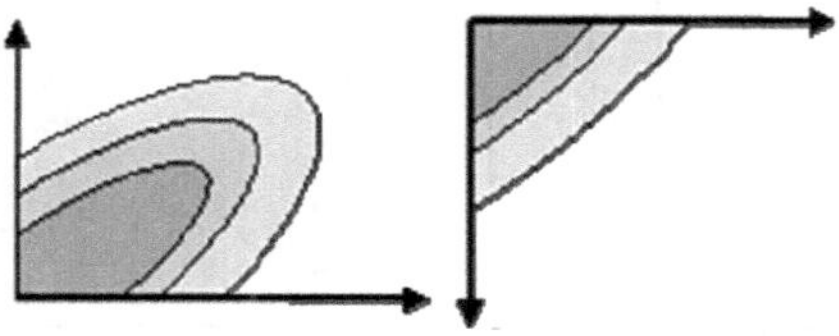

Fig. 7.6 Schematic representation of the bivariate densities of X_1 and X_2 (*left*) and of X_2 and X_3 (*right*)

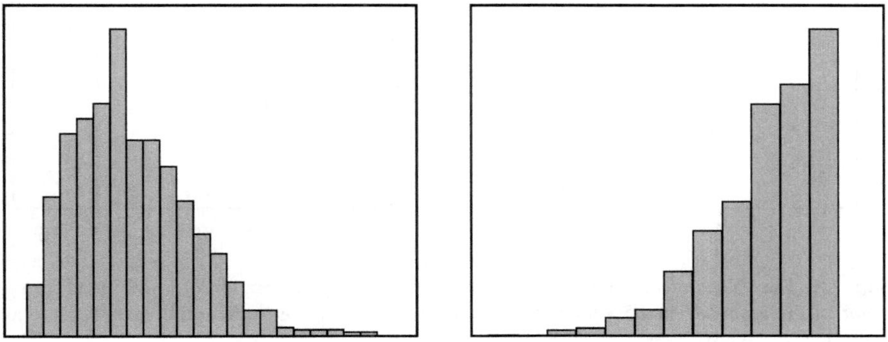

Fig. 7.7 The marginal distributions of X_1 and of X_3

of truncations is evident from these. In one case, we are dealing with the elongated part of the ellipse whereas in the other, it is merely a triangular "corner". Intuition can often be misleading when trying to guess the properties of truncated gaussian distributions. For example many people expect the marginal distributions in this example to be "half gaussians". The marginal distributions given in Fig. 7.7 show just how wrong intuition can be and also confirms what was shown in the two variable case given at the beginning of the chapter.

The Impact of Different Parallel Runs

Up till now we have illustrated the difference between the burn-in period and the stationary part by focussing on individual components. An alternative is to run

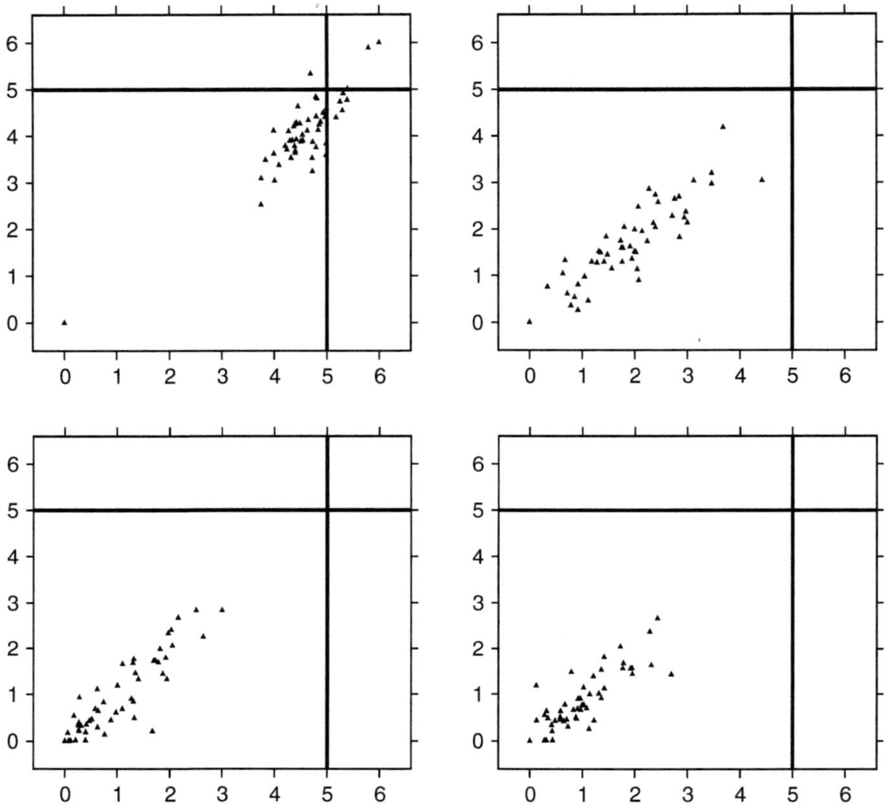

Fig. 7.8 The values of the first and second components after 1 iteration (*top left*), then 5 (*top right*), 10 (*bottom left*) and 50 iterations (*bottom right*) starting out from initial values of +5 (as indicated by the crosshairs)

many Gibbs samplers in parallel and study the results after a certain number of iterations. Figure 7.8 plots the first and second components for 50 parallel runs. Figure 7.8a shows their locations after a single iteration; both started out from an initial value of +5. Figures 7.8b–d give the output after 5, 10 and 50 iterations. As expected, the centre of the cloud moves downwards and disperses outward from this. Initially the distribution is far from the target cloud but as the number of iterations increases, it steadily tends toward it.

Chapter 8
Case Studies and Practical Examples

Choosing Which Simulation Method to Use

Simulation methods fall into two broad classes: pixel methods and object-based methods. Before presenting any practical examples, it is important to decide which method would be the most appropriate for the problem under study. So we first review the pros and cons of two classes of models, sequence-based pixel methods such as the truncated gaussian and plurigaussian simulations, and object-based methods, in particular boolean simulations. As this book focuses on plurigaussian simulations, we do not intend to present boolean simulations here. Interested readers can consult Matheron (1968, 1975), Lantuéjoul (1997a, b, 2002) Chilès and Delfiner (1999), and Molchanov (1997). Sequence-based pixel methods and boolean simulations can also be combined to get the best of both approaches. These are called nested simulations.

Sequence-Based Pixel Models

As their name suggests, sequence-based models are well suited to simulating geological configurations where lithotypes are organised in sedimentary sequences. Within a sedimentary sequence, lithotypes are correlated both vertically and horizontally. According to Walther's law, vertical successions of lithotypes also imply a lateral correlation between the lithotypes. Another advantage of pixel simulations is that they realistically reproduce the facies interfingering. Sedimentary sequences occur at different scales, and their organisation is a function both of auto and allocyclic processes. For example, fluvial channel sequences are characterised by a particular succession of lithotypes from base to top as the channels filled. The progradation of a shoreface which is mostly controlled by relative sea-level variations also shows a typical vertical succession of facies from offshore to foreshore. Over the past fifteen years, tremendous progress has been made in sequence stratigraphy and this has led to a much better understanding of the facies architecture of depositional sequences of different orders, in a wide range of depositional environments.

M. Armstrong et al., *Plurigaussian Simulations in Geosciences*,
DOI 10.1007/978-3-642-19607-2_8, © Springer-Verlag Berlin Heidelberg 2011

Even though the relative volume of facies and their geometric organisation (the facies partitioning) changes according to sequence organisation and although the sequences are usually non-stationary, proportion curves have proved to be a simple but highly effective tool for imaging sequences and for quantifying the associated non-stationarity. As they measure the relative proportion of lithofacies, both vertically and horizontally, they provide a good way of describing the spatial distribution of the lithofacies. They can guide the geologist in his interpretation and can help to define the reservoir layering (Eschard et al. 1999).

Vertical proportion curves give information on the facies partitioning within the sequences and indicate the stratigraphic levels where the main vertical permeability barriers can be expected, which is important in reservoir characterisation.

Computing horizontal proportion curves also helps in understanding the spatial distribution of the lithofacies and the geometry of the reservoir. Their representativity depends on the well spacing and distribution in the area to be simulated. When calculated in different directions across the field, they can show whether there are any major lateral facies variations in the study area. Furthermore, they can aid in deciding whether there is non-stationarity. When the relative proportions of facies vary significantly laterally within a reservoir unit, then non-stationarity is present.

In sequence-based pixel models, the correlation between facies is quantified by the vertical and horizontal variograms. In standard geostatistical studies, the range of the variogram has a direct geological significance; here the relationship is indirect. It is a function of the size of the heterogeneity, but also of its spatial frequency. The type of variogram used in the simulations also has a strong impact on the way the simulated facies are organised. Exponential variograms, for example, are appropriate when facies transition is progressive. On the contrary, gaussian variograms generate more "rounded" shapes in simulations.

Many depositional systems can be simulated well by sequence-based pixel approaches. They can handle carbonate depositional systems, which are often characterised by a cyclic organisation of facies and progressive facies interfingerings. In clastic depositional environments, the approach is also suitable for deltaic or shoreface sediments which are organised in transgressive-regressive cycles. Internal heterogeneity within fluvial sand sheets can also be simulated in this way. These types of reservoirs often show an internal sequential organisation which can be reproduced easily by sequence-based models.

One difficulty in the method is computing reliable horizontal variograms from wells because the well spacing is generally greater than the mean heterogeneity size. Different variogram ranges can then be tested, and the one which provides the best fit with production data is chosen (Eschard et al. 1998). Alternatively, data bases on analogous reservoirs or outcrops can be used to give an idea of the range.

Object-Based Models

Object-based models were designed to simulate geological objects which have a well-defined geometry. These generally correspond to sedimentary bodies isolated in a non-reservoir matrix (e.g. fluvial channels in floodplain mudstones). Similarly, shale breaks with a specific geometry within a massive sandstone unit can also be generated as objects. Different types of objects can be simulated together. Attraction or repulsion functions can also be applied to promote or prevent connections between different types of objects. Objects also can be superposed with erosion or preservation rules. The vertical frequency of the objects can be computed at different levels from the well data.

Generally speaking, object-based simulations require a great deal of *a priori* geological information. First, the geologist has to determine and classify the different types of objects from the well data. Then, he has to impose a simplified shape on them. This requires a good knowledge of the geological setting, and also the possibility of consulting reservoir data bases containing the geometry (width versus thickness, etc) and shapes (sinuosity, etc) for each depositional environment. This information generally comes from outcrop studies, from observations of comparable modern depositional environments or from well-documented subsurface fields. It is generally summarised in the form of cross plots and frequency diagrams in which the variability of the measured parameters is shown. This variability is often high for a given sedimentary object because the shapes and the amalgamation rate of the objects depend on their location within a sequence (Eschard et al. 1999). During periods when there is little space available for sedimentation, channels deeply incise the floodplain mudstones, and the amalgamation rate of fluvial channels is high. When this space is large, the floodplain aggrades, and channels become less incised in floodplain mudstones. Although this type of information is critical for controlling the object size, it is rarely available in reservoir data bases.

Siliclastic reservoirs can often be simulated by object based methods. The most classical objects that are simulated with this approach are fluvial channels and crevasse splays in a matrix composed of floodplain mudstones. Channels can either be straight or sinuous, isolated in the floodplain or amalgamated. Crevasse splays are connected to the channels.

Nested Simulations

Nested simulations combine object and sequence-based pixel models. Typically, the objects are simulated first, then filled with lithotypes using a sequence-based simulation. This approach is generally used to simulate large geological objects which present internal heterogeneities. For example, the heterogeneity in fluvial and estuarine channels corresponds to mudstone plugs deposited within the channels. The object, the channel, is simulated first, then filled by heterogeneous material using a sequence-based algorithm.

A similar approach was used by (Cozzi et al. 2002; Felletti 2004) for simulating turbidites. Rather than using objects to define the outer boundary of the turbidite units he kriged the exterior contours and then used mono-gaussians and factorised exponential variograms.

Building up the Reservoir or Orebody Model

The first step in a plurigaussian simulation of an orebody or a reservoir is to calculate the proportions of each facies. But even before doing this, we have to make a series of preliminary choices. We have to define the lithotypes to be studied, to divide the reservoir or orebody into units, to choose the references levels to be used for flattening each unit and finally to choose the parameters of the grid to be simulated.

Step 1: Defining the Lithotypes

The lithotypes constituting the orebody or reservoir have to be defined from the available data: core samples, well logs and seismic data. They have to honour the geological information, and where applicable, the petrophysical information. In general, the geologists define more facies than can reasonably be simulated. So these have to be grouped into lithotypes.

The definition adopted is crucial, firstly in the data integration process and later in the simulations themselves. For example, as well as reproducing the geology of the reservoir realistically, reservoir simulations must reproduce the key reservoir properties (porosity, relative permeability, capillary pressure, etc) from a fluid flow point of view. This implies reconciling the geological and petrophysical descriptions of the reservoir when the lithotypes are first defined. Furthermore, the lithotype data base must be homogeneous in all the wells.

Step 2: Dividing Reservoir or Orebody into Units

Reservoirs generally consist of several stacked units. Their geology may differ from one unit to another. So each unit must be simulated independently with different parameters. Variogram ranges and anisotropies are generally different for each of the units. Sometimes the geostatistical techniques used can vary from one unit to another, depending on their geology. It is not possible to correlate lithotypes across unit boundaries, and moreover the simulation grid follows the geometry of the units. This is why the definition of the reservoir units is a such critical step in reservoir modelling.

Step 3: Defining the Reference Level

The reference level for the simulation is a specific geological layer which is used to restore the geometry of the orebody or reservoir at the time of deposition. This level must have been deposited horizontally during sedimentation and should, if possible, correspond to a time line. The reservoir is then flattened using this as the reference level. Different reference levels can be used for each reservoir unit, or one can serve for several of them.

As for the definition of the lithotypes, the choice of the reference level has a marked impact on the result of the simulation. For example, onlap or toplap configurations can be produced depending on the choice of a reference level with regards to the unit geometry (Fig. 8.1).

Step 4: Choosing the Grid Spacing

The grid size is in theory directly dependent on the size of the heterogeneities to be reproduced in the simulation. The detail needed in a simulation also depends on the recovery process used to produce the reservoir. Gas is less sensitive than oil to small scale reservoir heterogeneities. Fields developed with horizontal wells generally require very detailed reservoir models. From a geological point of view, it is tempting to simulate as much detail as possible, but in practice, the number of cells is limited by computer capacity. Because of the limitations on fluid flow simulators, the fine grid simulated has to be upscaled to a coarser one. Within each cell of the latter grid, the upscaling algorithm has to summarise all the values in the fine grid in a single value. This is quite simple for porosity where it means taking the average, but it is more difficult for permeability. Only approximate solutions exist (unless we run a fluid flow simulator for each cell of the coarse grid, and that would be prohibitively time consuming). Consequently many different upscaling techniques exist and so differences may appear in the simulation depending on which upscaling method is applied, especially when the number of cells is very large.

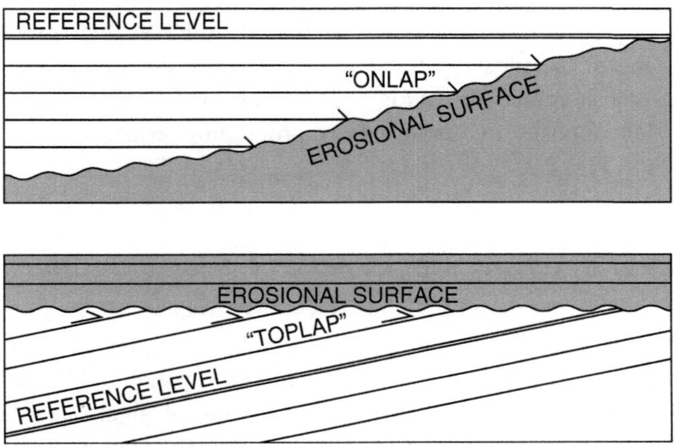

Fig. 8.1 Onlap (above) or toplap (below) configurations are produced depending on the choice of a reference level relative to the unit geometry

Petroleum Applications of the Plurigaussian Approach

The geostatistical methods presented above have their limitations when applied in complex geological settings. In some cases, the reservoir architecture presents complex facies transitions which cannot be simulated with mono-gaussian techniques. Similarly, boolean models are only suitable when only a few types of sedimentary bodies are present and when their shapes are well defined. The main advantage of the plurigaussian approach is its ability to handle complex facies relationships, both vertically and laterally, in a pixel simulation. For example, it is possible to impose different anisotropies onto the gaussian functions. Furthermore, different types of variograms each with its own range and anisotropy can be used for each gaussian function. The combination of all these parameters makes the approach very flexible.

Constructing the Lithotype Rule with Geological Constraints

As the theory of the lithotype rule construction has already been described, we will illustrate it with several examples to show how the rules can be chosen *a priori* depending on the geological context. As usual, lithotypes are represented by different colours and the surface occupied by each one in the matrix is a function of the relative proportion of facies computed in wells. Complex facies transitions can then be reproduced by changing the relative position and surface of the lithotype in the matrix.

If only one gaussian function is used, the lithotype rule only shows superposed bands, with each band representing a lithofacies. When colours are in contact in the lithotype rule, the corresponding lithotypes will be connected in the simulation. Consequently the ordering of the facies in the lithotype rule must then respect a sedimentary sequence in order to correctly reproduce facies inter-fingerings in the simulation. One of the most common sequences encountered corresponds to the progradation of a shoreface (top of Fig. 8.2), in which offshore, shoreface, foreshore and coastal plain deposits are organised in a prograding sequence. Its lithotype rule is also shown in Fig. 8.2.

Simulation of Reservoir with Complex Facies Transitions

The lithotype rule also makes it possible to simulate complex geological settings, especially in carbonate depositional systems. For example, it is difficult to simulate the irregular shape and complex internal architecture of algal or reefal bioherms with classical object or sequence based models (Van Buchem et al. 2000). In this case, the lithotype rule is used together with the vertical proportion curve as a kind of facies substitution diagram to reproduce the sequential organisation of the facies.

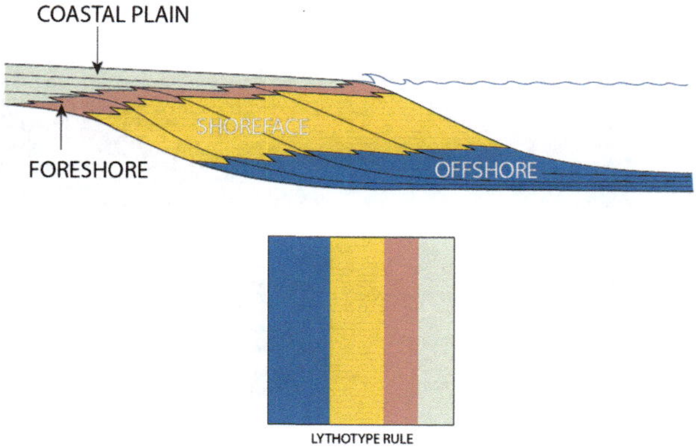

Fig. 8.2 Lithotype rule with only one gaussian function used to simulate a prograding sequence. Facies ordering is respected in the rule

Figure 8.3 shows algal mounds in the Paradox basin, of Pennsylvanian age, which were studied in outcrops along the San Juan River (Grammer et al. 2000; Van Buchem et al. 2000). The algal bioherms correspond to bioconstructions with irregular rounded shapes which were constructed in a mixed carbonate and siliciclastic open shelf. The carbonate shelf first prograded (facies 1–3), and then when a certain water depth was reached, incipient mounds (facies 4) started to develop in the shelf setting.

After this initial stage, the mounds themselves started to grow, and different stages of construction have been observed (facies 5: initial stage, facies 6: final stage). At the end, the mounds have an irregular rounded shape with a height of more than 20 m compared to the surrounding shelf. When the algal mound growth stopped, the inter-mounds troughs where progressively filled by in situ shelf sediments and material falling from the mounds. These deposits now form flanking beds in outcrops (facies 7), with the beds onlapping the mound relief.

Finally, clastic deposits (facies 8) were deposited in the shelf setting during subsequent relative sea-level drops. On top of the sequence, karstification often restricted marine carbonates (facies 9) during emersion.

The lithotype rule corresponding to this complex geological setting is shown in Fig. 8.4. Firstly, the series was divided in two units: the lower unit corresponds to the platform progradation, whereas the algal mounds are located in the upper unit. The lithotype rule of the lower unit is simple as only one gaussian function is used for its simulation. Its vertical proportion curve also reflects the progradational pattern of the shelf.

In the upper unit, the lithotype rule is more complex because of the spatial organisation of the mounds and intermound facies. The intermound facies (facies 7) can be in contact with the two mounds facies (facies 5 and 6) and with the sandstones capping both the mounds and the inter-mounds deposits (facies 8). The simulation in Fig. 8.5 reproduces the geometry of the mounds in this shallow marine platform setting, very realistically both in section and in plan view.

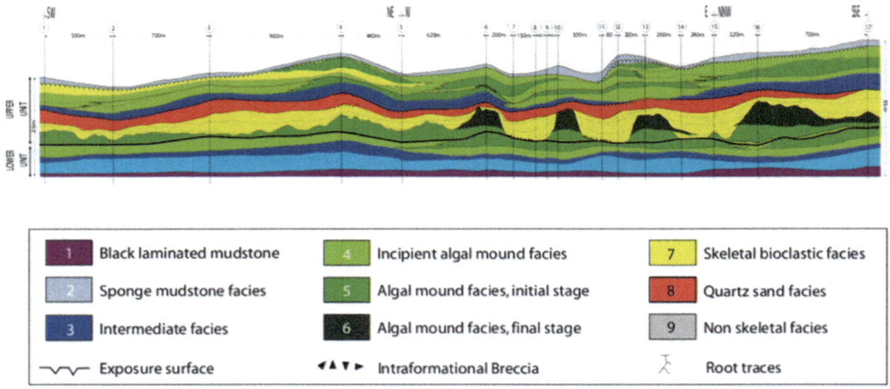

Fig. 8.3 Geological model of Pennsylvanian algal mounds in outcrops (from Galli et al. 2006)

Fig. 8.4 Lithotype rules for the upper (on the right) and lower units of Paradox basin

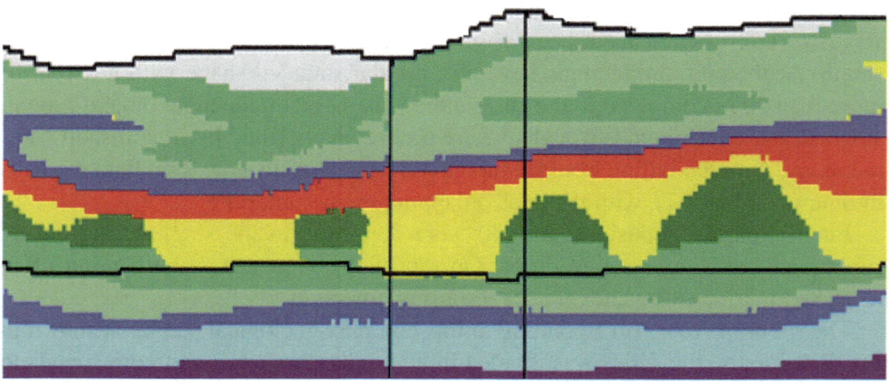

Fig. 8.5 Cross section in a simulation of the Pennsylvanian algal mounds using the pluri-gaussian approach. The mound geometry is realistically reproduced (compare with the geological model in Fig. 8.3)

Plurigaussian simulations are particularly suitable for simulating interdependent phenomena which interacted during sedimentation or just afterwards. Moreover, the two phenomena can have different anisotropies, and can be correlated or not.

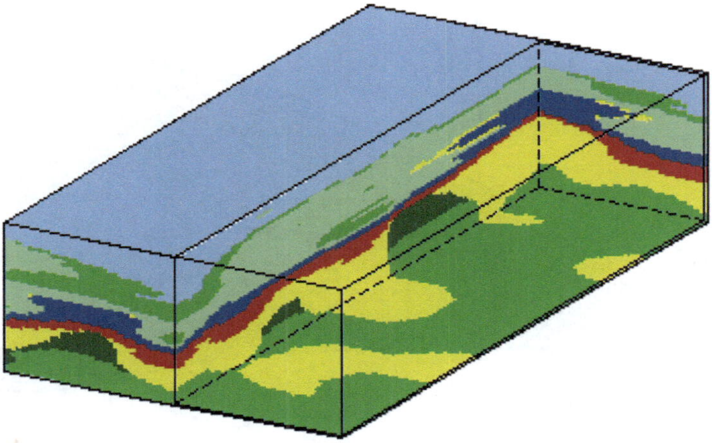

Fig. 8.6 Simulation showing the altered facies capping the mounds. The altered facies corresponds to the *light grey* colour on top of the dark algal mound

The effects of the primary diagenesis, which affect the sediments just after burial, can be simulated with this approach.

The Paradox basin algal mounds described earlier are a good example of the effects of primary diagenesis. During relative sea-level drops, the topmost part of the mounds was altered when the mound top is emerged. This alteration is a function of the initial carbonate texture and porosity. The flanking beds which were deposited after this episode, were not affected. A new facies corresponding to this alteration of the reef was added into the lithotype rule (facies 9). It is in contact with both the flanking beds laterally, the mound core below and the sandstones above. The simulation in Fig. 8.6 reproduces the altered facies capping the mounds.

Simulation of the Effects of Primary Diagenesis in Complex Reservoir

A synthetic case study of a reef reservoir was carried out in order to test the ability of the plurigaussian method to simulate this type of reservoir. It is based on the Miocene reefs observed in subsurface by Gröstch and Mercadier (1999). In this synthetic case (Fig. 8.7), the reef growth occurs in three stages. Each stage is characterised by a specific organisation of the facies. The basal unit (unit III) corresponds to the initial stage of the reef growth when no lagoon had developed inside the reef. The middle unit (unit II) corresponds to the growth stage of the reef. Back-reef and lagunal deposits were well developed inside the reef. The topmost layers of unit II were cemented during a relative sea-level drop when the topmost part of the reef emerged. Unit I is the final stage of the reef growth. According to

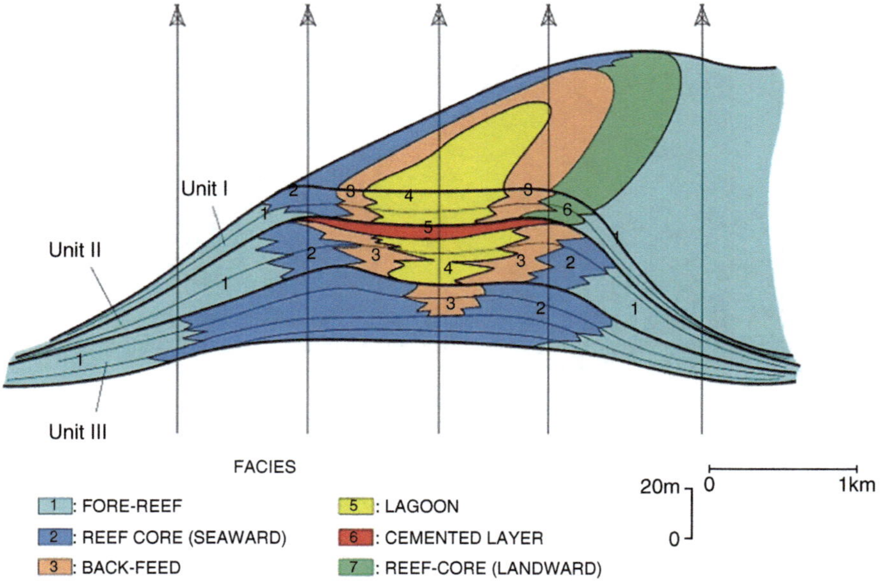

Fig. 8.7 Synthetic model of a reef complex from which a pseudo-well data-base was compiled for simulation purposes

Grötsch and Mercadier (1999), cementation can specifically affect the core reef exposed to the ocean but not the lee-side where porosity is preserved.

Six facies were differentiated for the simulation: fore reef (facies 1), core reef, basin-ward (facies 2), back-reef (facies 3), lagoon (facies 4), cemented facies (facies 5), and core-reef landward (facies 6). Facies 5 corresponds to a cemented horizon which affected facies 2, 3 and 4 in unit II. The cemented horizon constitutes a significant permeability barrier within the reservoir, and so it is important to reproduce it in the simulation. Three litho-stratigraphic units were then used for the reservoir simulation, each with a specific facies distribution. The simulation was processed in a proportional grid within a unit, in order to get correlation lines parallel to the top and bottom of each unit.

In unit III, the truncated gaussian algorithm was used with a simple lithotype rule (Fig. 8.8). In unit II, the plurigaussian algorithm was applied in order to respect the complex relationship between the facies. The cemented facies (facies 5) can be in contact with facies 2, 3 and 4 but facies 5 cannot be in contact with facies 1. The vertical proportion curve also constrains facies 5 only to the topmost part of the unit II. In unit I, the porous core reef (facies 6) is only present in the lee side oriented towards the continent. The lithotype rule prevents contact between facies 6 and 4. Gaussian variograms were used for the two gaussian functions in order to respect the rounded shape of the reef facies belts. The continuity of the facies distribution was also ensured by using long variogram ranges of the 6,000 m in the East-West direction and 15,000 m in the North-South direction. This anisotropy imposes an ellipsoidal shape on the simulated reef. The results of the simulation are shown in

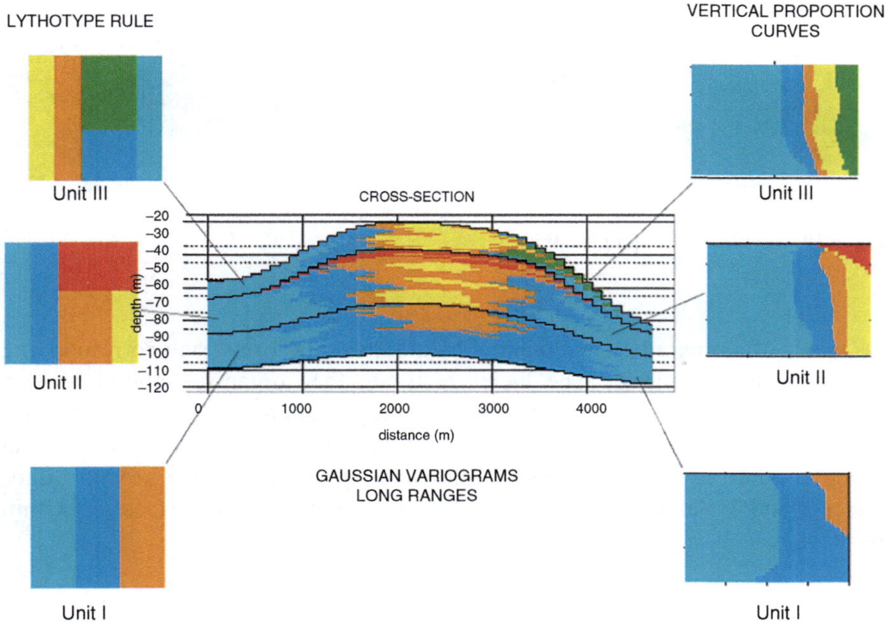

Fig. 8.8 Vertical section extracted from a plurigaussian simulation of the reefal complex which was divided into three units. These are shown with their lithotype rules (*left*) and their vertical proportion curves (*right*)

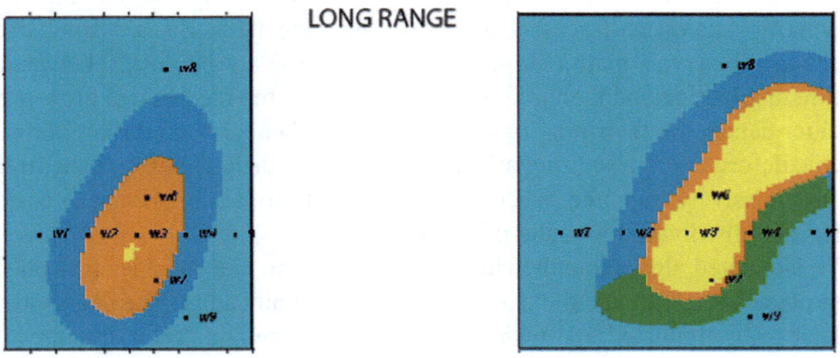

Fig. 8.9 Plan view of the reef simulation, using a gaussian variogram with a long range in Unit II (**a**) and III (**b**). See Fig. 8.8 for unit names

Fig. 8.8 in a vertical section across the reef, and in Fig. 8.9 in plan view. The complex internal organisation of the reef has been reproduced well in the simulation.

Simulating Progradational Patterns

In a progradational sequence, the facies are organised in a logical order from the deepest facies at the base of sequence to the shallowest on top. Laterally, the

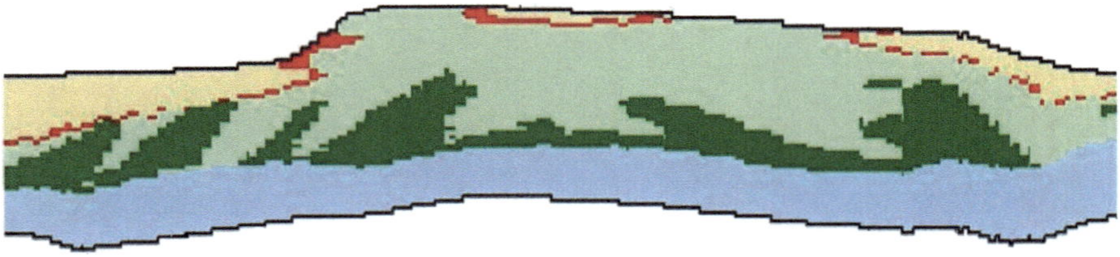

Fig. 8.10 Simulation of a progradation

geometry of a progradation is marked by oblique clinoforms each corresponding to a depositional surface The facies belts are then organised obliquely with their dip corresponding to the direction of progradation (Fig. 8.10). In it lithofacies 1 is separated from lithofacies 4 by lithofacies 2 and 3. The first gaussian function controls the thresholds of lithofacies 1 and 4, and the second, lithofacies 2 and 3. An anisotropy was added to simulate the dip of facies 3 within facies 4.

Simulation of Fractures Affecting a Specific Facies

Fracturing which affects some reservoirs after burial can also be simulated with a stochastic approach (Cacas et al. 2001). Fracture networks are simulated as objects in a reservoir matrix. In many cases, there is a relationship between the fracture distribution and the initial lithotypes because the mechanical properties of the latter depend on their lithology and textures. When stressed during burial and tectonic deformation, fracturing affects the lithotypes differentially. Some of them end up intensively fractured, whereas others do not. So it is important to correlate the fracture distribution with the lithotype.

We simulated algal mounds with fractures in them. We consider that only the mound facies (facies 5 and 6) have been fractured significantly. The distribution of the facies was simulated first with the plurigaussian method. Then, fractures were only simulated in the mound facies. Figure 8.11 presents the results.

Testing the Impact of Simulation Parameters

Different tests were carried out to evaluate the sensitivity of the simulations to variations in input parameters. The synthetic reef case study presented in Fig. 8.7 was used as a reference to evaluate the effect of changing the type of variogram model and its range. The influence of the range was tested on unit II. In the reference simulation

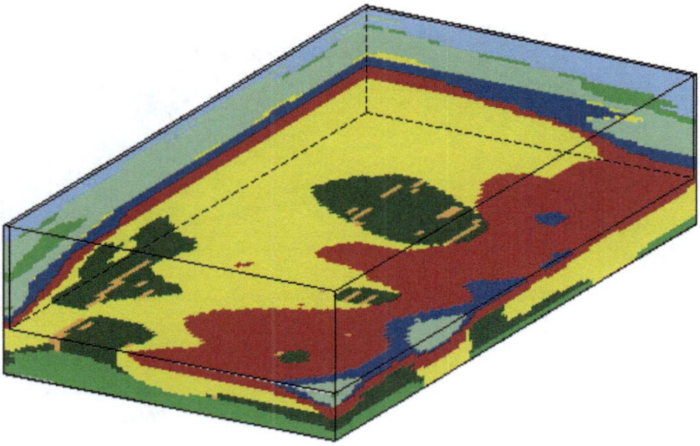

Fig. 8.11 Simulation of fractures only affecting one facies within the Pennsylvanian algal mounds

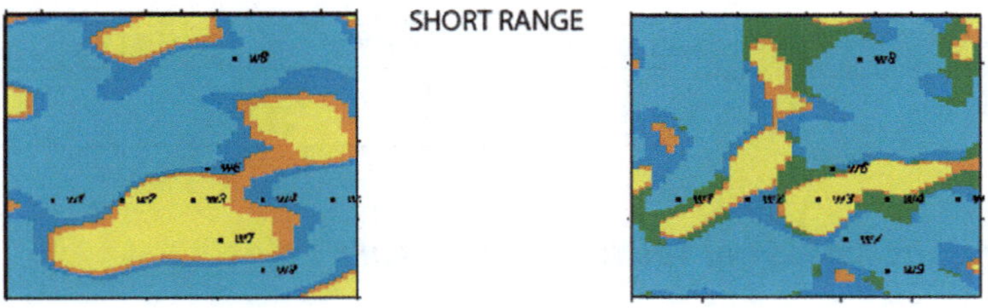

Fig. 8.12 Testing the variogram range influence on the reefal complex described in Fig. 8.7. The figure shows a plan view of the reef using variograms with short ranges. Compare with Fig. 8.9 in which the same plan was simulated with a long range

(Figs. 8.8 and 8.9), the horizontal ranges used were 6,000 m and 15,000 m in the X and Y directions respectively. In Fig. 8.12, smaller ranges were used (X = 1,500 m, Y = 1,300 m).

Comparing the two simulations shows that the range reduction has dramatic consequences on the shape and distribution of the facies belts. Small isolated reefs are produced by the smaller ranges. The sensitivity of the simulations to the variogram model was then tested. In the reference simulation, two gaussian variograms with ranges of X = 6,000 m, Y = 15,000 m, Z = 5 m were used for each gaussian function. Figure 8.13 shows the results produced using exponential variograms with similar ranges to the reference simulation. The consequences of this change are again very important.

The facies are more scattered in the simulation using an exponential variogram models than in the one using gaussian variograms. So the type of variogram model can have important consequences when computing facies connectivity within a reservoir.

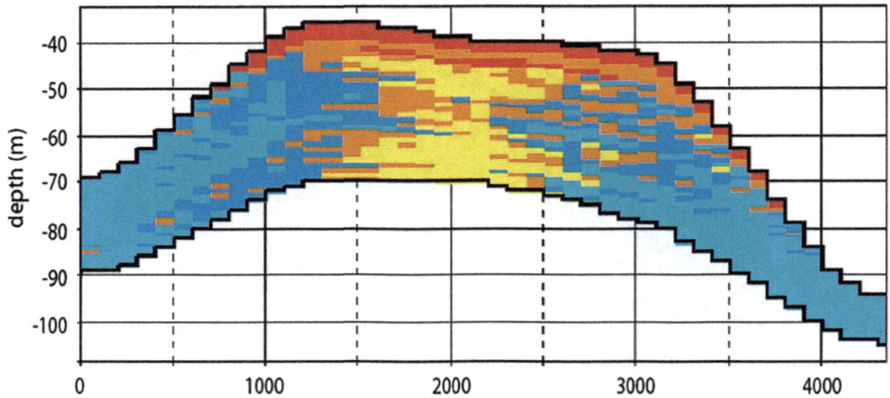

Fig. 8.13 Testing the variogram model. The simulation represents a cross section of the reef complex described in Fig. 8.7. An exponential variogram model was used here. Compare with Fig. 8.8 in which the same reef complex was simulated with a gaussian variogram

Table 8.1 Percentage of connected volume in one zone of the reservoir

Rock type	P90 (%)	P50 (%)	P10 (%)
N° 1	15	20	25
N° 2	27	31	35

Testing the Impact of the Rock-Type Rule

Normando et al. (2005) tested the effect of slightly changing the rock-type rule on the connectivity of the reservoir while fixing the other parameters. In a study on a mature Brazilian field with four facies, the authors showed permuting the position of two of the facies in the rock-type rule led to significant changes in the connectivity in simulated reservoirs obtained using the two different rules (Table 8.1).

Handling Heterotopic Data

The aim of the project between IFP, Ecole des Mines de Paris and the Italian oil company, ENI, was to tackle the problem of simultaneously simulating sedimentary facies and diagenesis index in a reservoir deposited in a clastic fluvial environment. Four sedimentary facies were defined from cores and logs of 15 wells. In addition four diagenetic indices were identified from thin section analyses for some of the wells. As the two properties are not known at the same locations, this dataset is heterotopic. This study gave us the opportunity to test different simulation approaches to produce a reservoir model including the two properties, and to

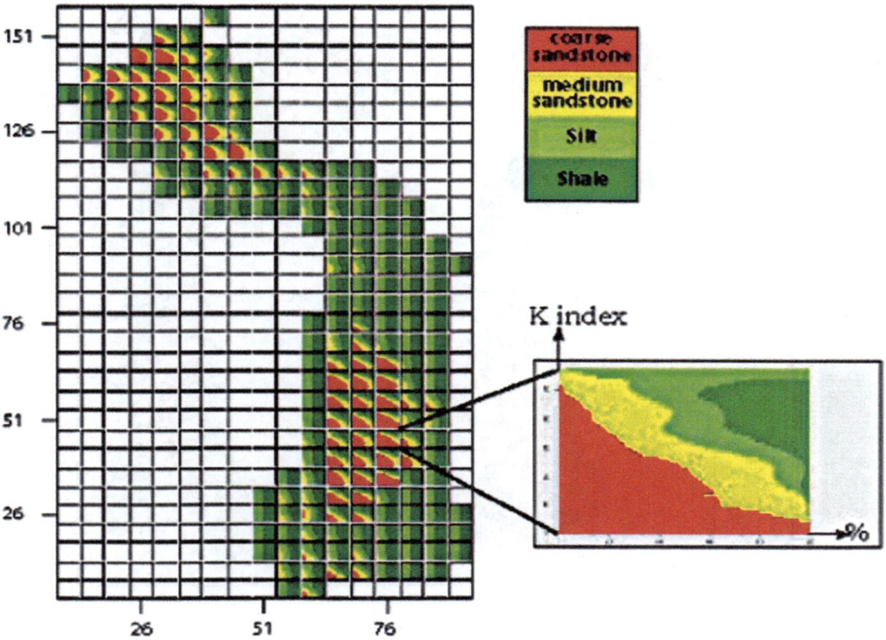

Fig. 8.14 Matrix of percentage for the sedimentary facies

identify the different characteristics of these methods. Three different approaches were used to simulate the sedimentary facies and the diagenetic

- Nested truncated gaussian simulations
- Classical plurigaussian simulations (PGS)
- Bi-plurigaussian simulations (Bi-PGS)

In the workflow for the nested simulations, the steps are performed sequentially: first a classical plurigaussian simulation of sedimentary facies, using a vertical proportion matrix in terms of sedimentary facies (Fig. 8.14); second, for each sedimentary facies, a classical plurigaussian simulation of the diagenesis index, and third the reconstitution of the final information using logical rules to combine the different realizations (Fig. 8.15). The simulation parameters are specific to each property, sedimentary facies and diagenesis index. This method makes it possible to separate the constraints for each simulated property in terms of conditioning wells and parameters. The diagenetic property is independent from one cell to another, if the sedimentary facies change between these cells. The disadvantage is that it will not be possible to use any correlations found between the sedimentation and the diagenesis.

The classical plurigaussian simulation method (PGS) can be used directly if the two indicators variable are combined into one single global indicator variable, called the facies-diagenesis variable. In that case only the wells where both types

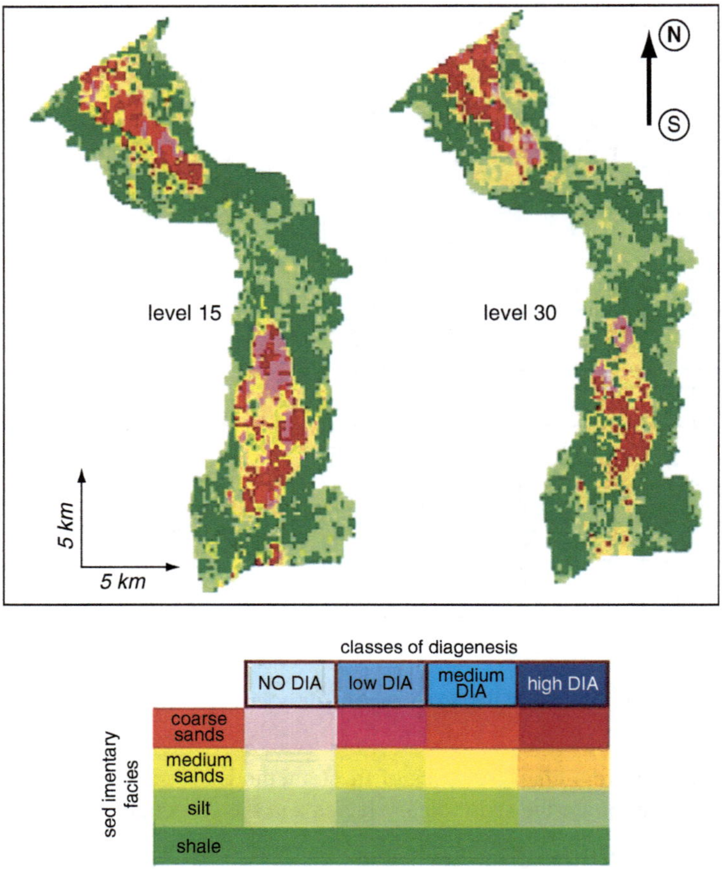

Fig. 8.15 Two levels in the reservoir model using nested truncated Gaussian simulations

of information, sedimentary facies and diagenesis index, are defined can be used for the conditioning.

A single simulation where each Gaussian function is associated with one indicator variable, (here G1 is related to the sedimentation and G2 is related to the diagenesis) is carried out globally. A correlation coefficient between the two Gaussian functions could be used to take account of a relationship between sedimentation and diagenesis (Fig. 8.16). This approach is in fact a short-cut to combine in one simulation two mono-truncated Gaussian simulations. Thus the continuity in the diagenesis property through the facies is respected by construction.

Fig. 8.17 illustrates the main differences between the three simulation methods. The bi-plurigaussian simulation (Bi-PGS) model has been also used as it provides a sound basis for bivariate categorical simulation. In this approach, each physical process is associated with a complete PGS. The two vertical proportion matrices

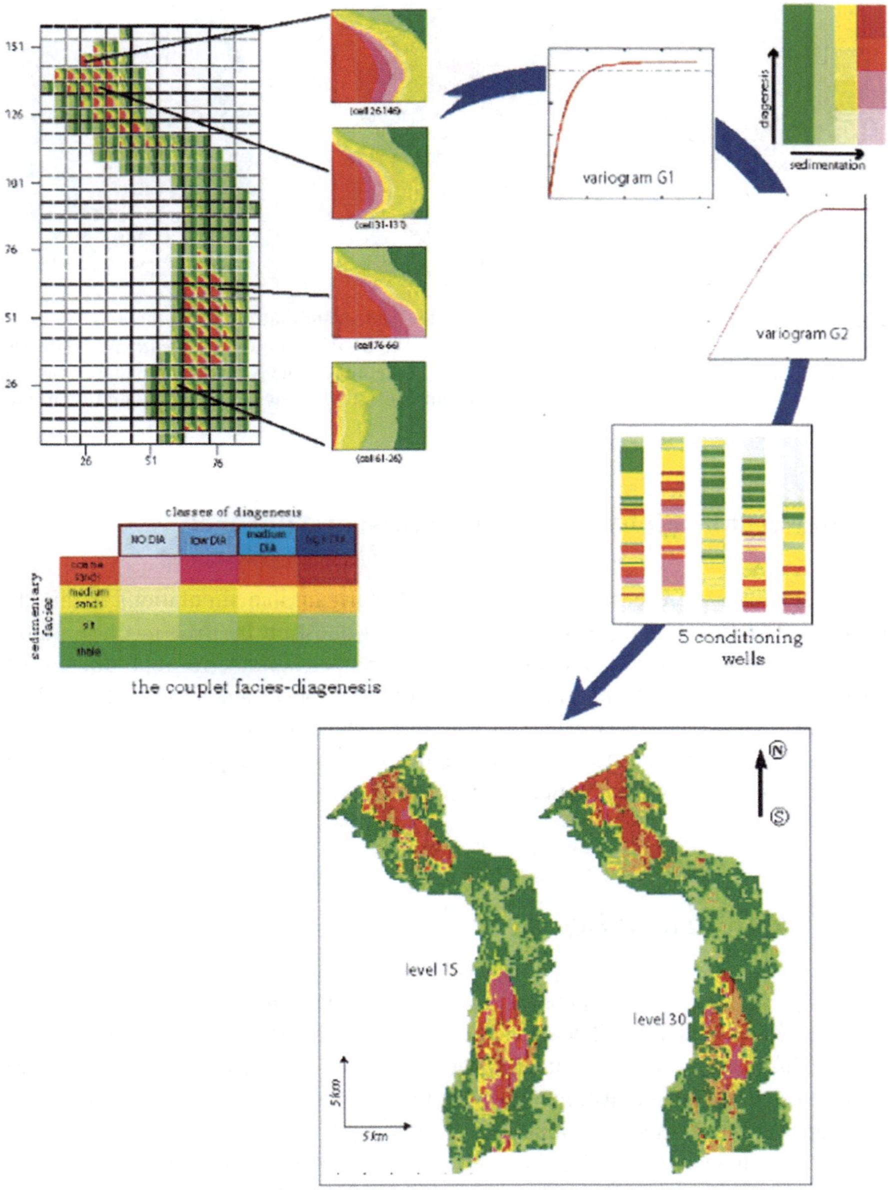

Fig. 8.16 Workflow and parameters for plurigaussian simulation

are linked through the conditional probabilities between the two indicators. The advantage of this approach is that all the data are taken into account during the conditioning, even the ones where only one indicator is available (Fig. 8.17).

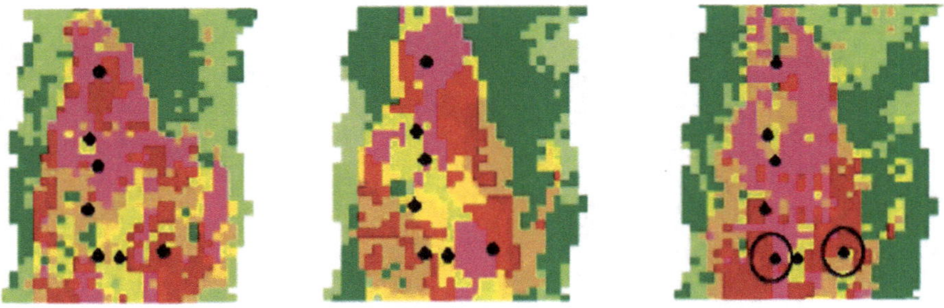

Fig. 8.17 Comparing the three approaches. The nested simulation (*left*) there is no continuity (correlation) in the diagenesis variable. In the Bi-PGS simulation (*centre*) all data are used and there is continuity in the diagenesis. In contrast the wells where there was no diaenesis data were not used the PGS (*right*) and so the wells with circles around them are not respected

Mining Applications of the Plurigaussian Approach

In this section we present two applications of plurigaussian simulations to mining projects. The first was developed in answer to a question by the French uranium mining company, Areva: Is the plurigaussian approach suitable for reproducing the characteristics of roll-front uranium deposits? As these deposits are formed in permeable sandstone, the method is a natural extension of the techniques in the oil industry. The second case-study is an application to a porphyry copper deposit in Chile, that is, in a non-sedimentary environment.

Roll-Front Uranium Deposit

In this application, we are interested in a roll-front uranium deposit and its exploitation by in-situ leaching. The dataset comes from the Muyumkum deposit in the South Khazakstan (Fontaine and Beucher 2006). Roll-front uranium deposits are formed in permeable sandstone. The uranium is located at the interface between oxidizing and reducing conditions.

The in-situ leaching technique consists of circulating oxidising and acid or alkaline solutions in the mineralised area via injection wells to dissolve the uranium selectively. As a consequence, there are two issues: on the one hand, to delimit the areas where the uranium grade is above a given threshold and on the other hand to reproduce the distribution of the hydrodynamic parameters to mimic the fluid flows for the in-situ leaching.

First Issue

As the hydrodynamic parameter distribution depends mainly on the geological characteristics, this constitutes the first step of the study. In the example, the facies have been grouped in three classes from silt to medium and coarse grained sands. Moreover the sands have been split into non-oxidized and oxidized facies. So, seven types were defined. The spatial distribution of the proportions (see vertical proportion curves, Fig. 8.18, and vertical proportion matrix, Fig. 8.19) shows that the oxidized types are predominantly in the south east which is correlated to the origin of the oxidizing fluids. Vertically, these facies are more abundant in the upper part of the deposit. The shale facies has almost the same proportion in the whole domain.

As these seven types result from two processes, two underlying Gaussians are used. The first one gives the granulometric evolution from coarse sand to silt and the second one separates oxidized and non oxidized types. The shapes of the experimental variograms confirm the choice of the lithotype rule. The fits of the oxidized facies is presented in the figure.

Giving this lithotype rule, the fit of shale type gives the parameters for the first gaussian random function. Then the second one is chosen by fitting the other experimental indicator variograms. The result is presented for the oxidized facies in Fig. 8.20. A vertical E-W cross-section in a conditional simulation (Fig. 8.21)

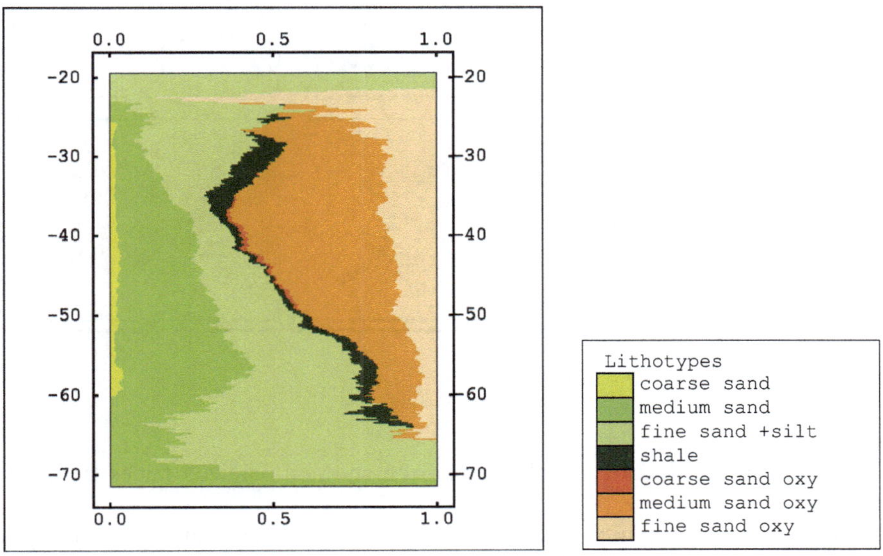

Fig. 8.18 Vertical proportion curves (VPC)

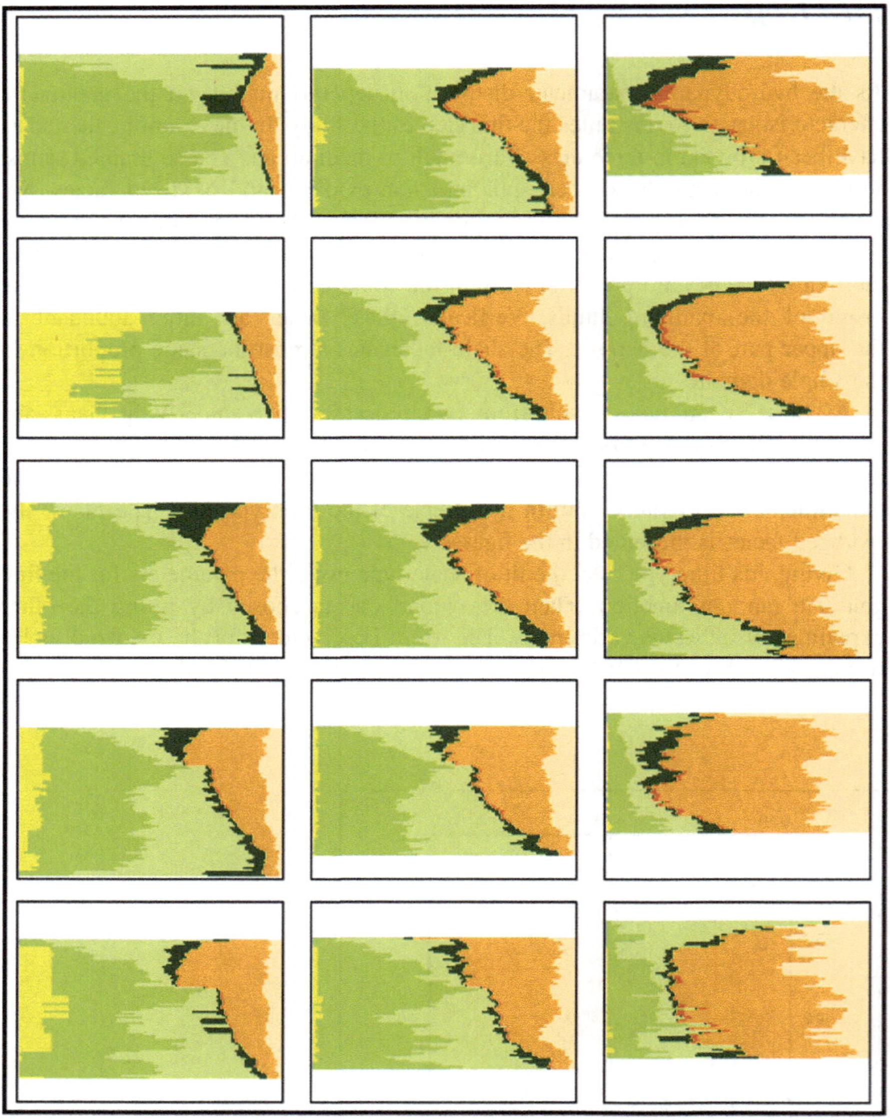

Fig. 8.19 Vertical proportion matrix (VPM)

shows the spatial distribution of the oxidized and non oxidized facies from east to west.

A 3D view of the simulated shale (Fig. 8.22) illustrates the relative stationarity of this facies in the field as visible in the vertical proportion matrix while Fig. 8.23 shows a 3D view of the simulated oxidized facies.

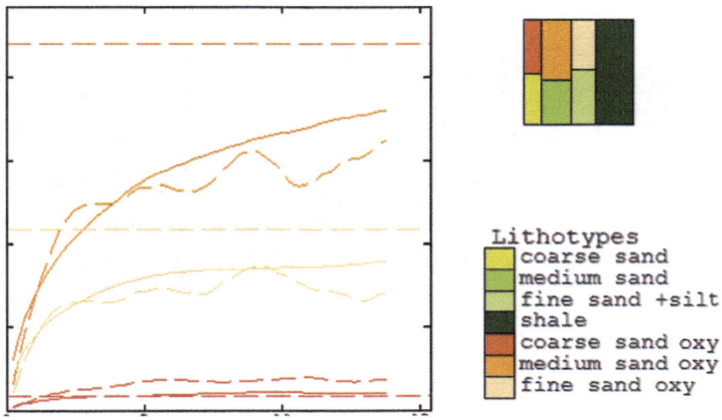

Fig. 8.20 Experimental and fitted vertical variograms for the oxidizing lithotypes

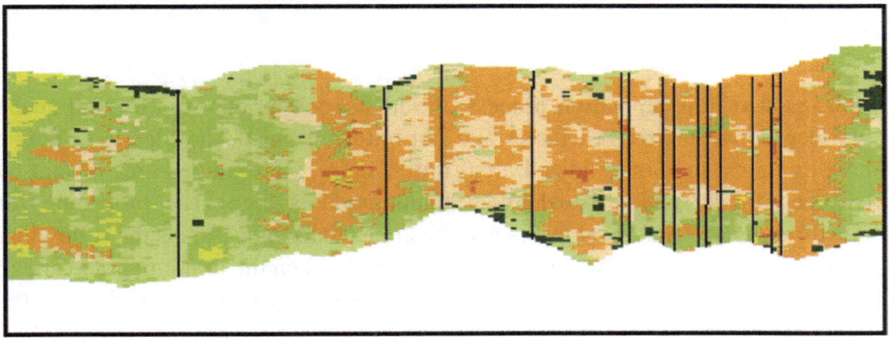

Fig. 8.21 Vertical cross-section on one of the simulations showing the seven lithotypes

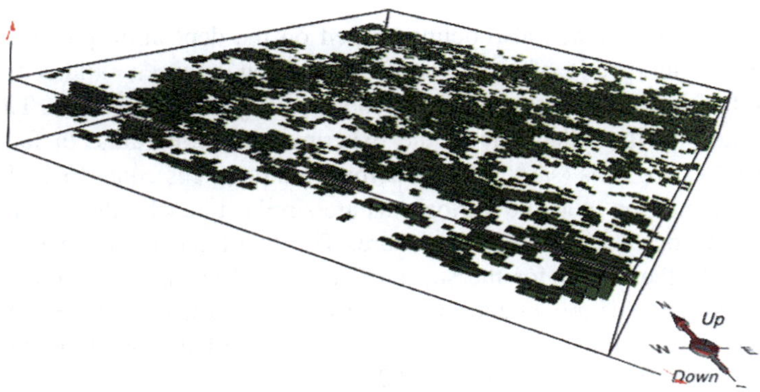

Fig. 8.22 3D block model of the simulated shale facies

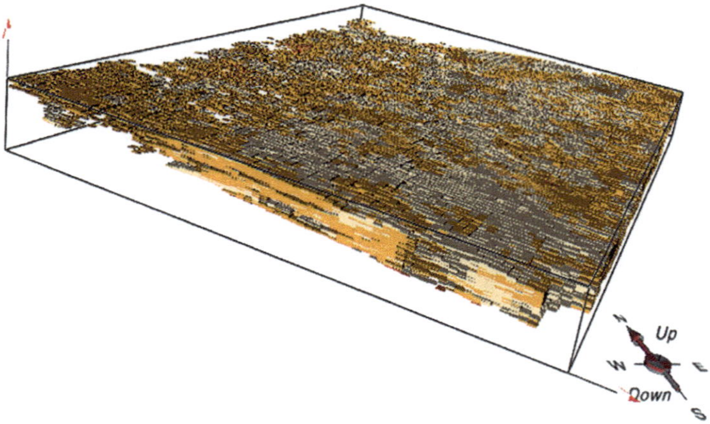

Fig. 8.23 3D block model of simulated oxidized facies

Second Issue

As a result of the deposition process, the high grade uranium is located at the interface between oxidized facies and not oxidized ones. The idea is to define three domains: the non oxidized area (initial state), the oxidized area (swept by oxidizing fluids) and area where uranium grade is high. The procedure for transforming the geological facies and the grades into indicators is the following: the high grade indicator is obtained using a threshold on the uranium grade and in the rest, the initial state and the oxidized one, is distinguished depending on the oxidation factor.

Globally the high grade ore is more or less constant along the vertical axis (see VPC Fig. 8.24). In the 3D space it is mainly located in the North West part along a West south- North east axis (see the 3D view of the high grade proportion upper than 30% (Fig. 8.25). This is the result of the deposition process.

The lithotype rule was also defined based on the deposition process: the first underlying Gaussian is affected to the distinction between oxidized and non oxidized facies while the second one depends on the grade information. This choice is confirmed by the behaviour of the experimental variograms. In fact, on the normalized variograms (same variances for the three types equal to 1) the shapes of oxidized and non oxidized variograms are similar to each other but are different from the one for the high grade ore. The fit of the high grade variogram gives directly the model for the second Gaussian. The fits of the other types is a combination of this model and the model for the first Gaussian. (Fig. 8.26) The conditional simulations were then performed with all the previous parameters. A typical simulation is shown in Fig. 8.27.

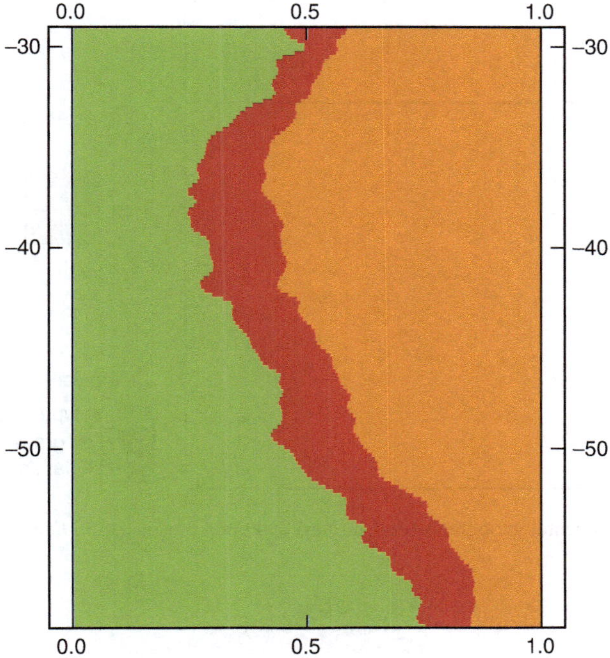

Fig. 8.24 Global vertical proportion curves for the three types

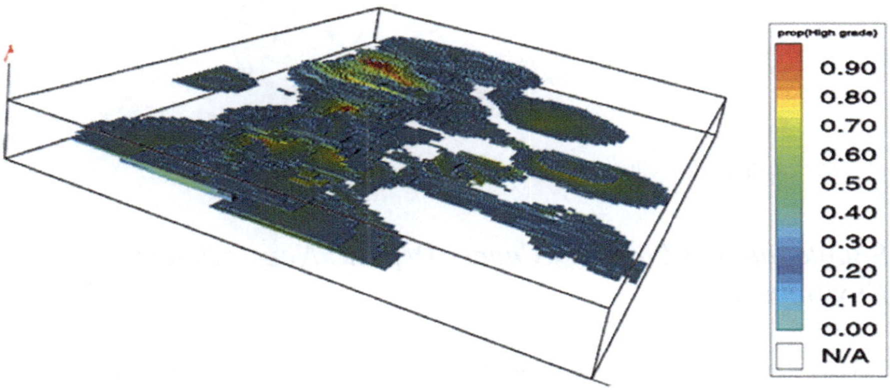

Fig. 8.25 Proportion of high grades above 30%

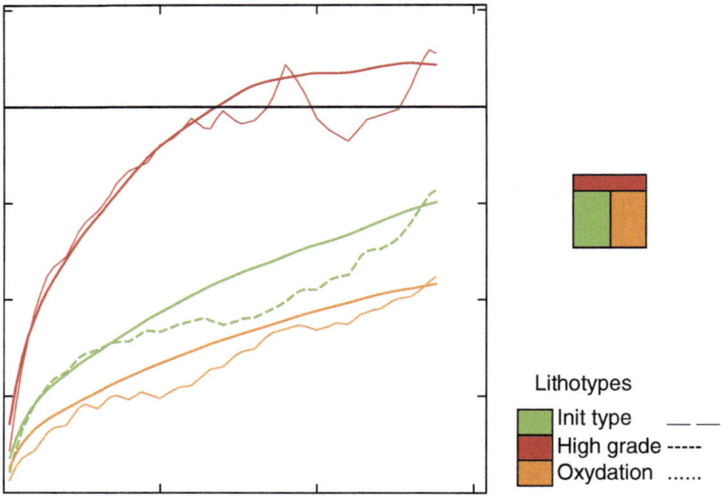

Fig. 8.26 Experimental variograms (*dotted lines*) and fitted models (*solid lines*)

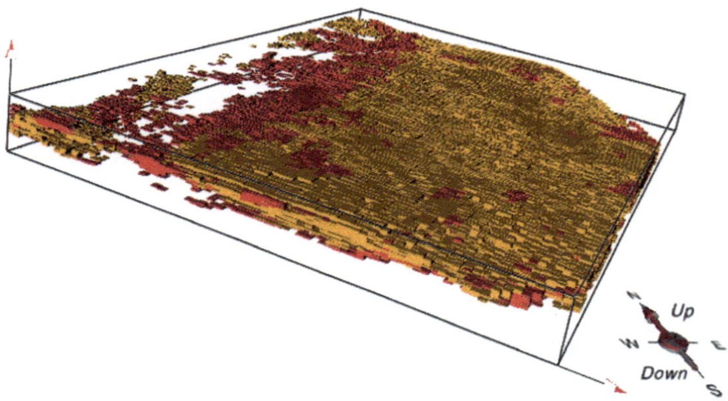

Fig. 8.27 A simulation with only oxidised and high grade facies shown

Simulation of a Porphyry Copper Deposit: Non-sedimentary Environment

This case-study has been included to show that plurigaussian simulations can also be used in non sedimentary geological environments. In these cases, there is no longer any reason to use vertical proportions, as the geology is not guided by the vertical sedimentation.

The MM copper deposit is a porphyry copper deposit near the Chuquicamata mine. In this type of deposit, the mineralization is constrained by rising fluids through a network of faults and joints and by the interactions of these fluids with the pre-existing minerals. In this orebody, this network is controlled by a major sub vertical North-South fault, called the MM fault. As the auxiliary faults and joints principally follow this fault, the copper mineralization, which controls the working rock types used for this study has a general North South vertical orientation. Four rock types have been defined, corresponding to different stages of copper enrichment: breccias (higher grade ore), C5 stockwork (also high grade ore), the low grade ore C1, and the waste rock. Breccias and C5 are generally in direct contact with the faults of joints network. The copper concentration of the fluids decreased as they went further from this network. Their interaction with the pre-existing rocks created the low grade rock type C1.

Further out, there was not enough copper left in the fluids to lead to mineralization and the resulting rock type is waste. Figure 8.28 shows the distribution of the rock types in an east-west vertical section.

As the rocktype depends mostly on the distance from the fault, it seems natural to use the fault plane which is oriented in a North-South vertical direction, as the reference level. The flattening consists in shifting the East-West coordinates to transform the fault into a perfect North-South vertical plane. Consequently, the

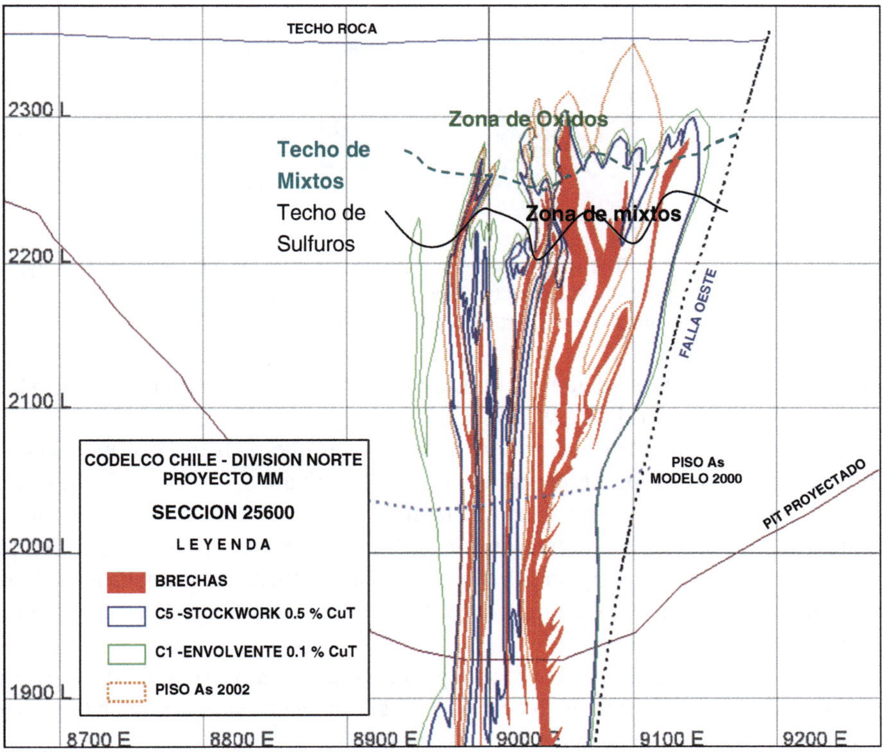

Fig. 8.28 Schematic distribution of the rocktypes. in a vertical east-west vertical section. Reproduced from Riquelme et al. (2008) with permission

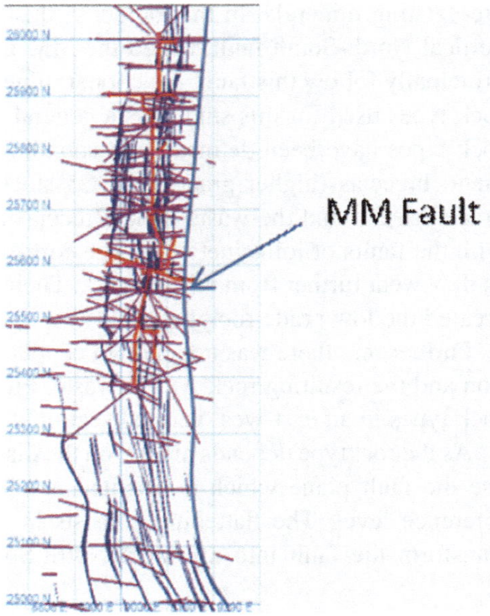

Fig. 8.29 Horizontal projection view of the MM fault, the breccias (*blue*) and the drillholes (*pink*). Reproduced from Riquelme et al. (2008) with permission

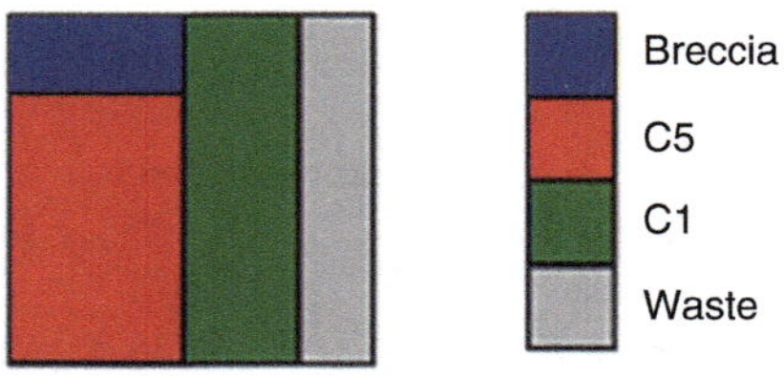

Fig. 8.30 Rock-type rule. The first gaussian separates C1 (*green*) from the richer rock types (breccias + the C5 group) on one side, and from the waste (*grey*) on the other side, while the second separates breccia (*blue*) and C5 (*red*). Reproduced from Riquelme et al. (2008) with permission

proportions will be plotted along an East-West direction, perpendicular to the MM fault in the working grid (Fig. 8.29).

The rock type distribution varies perpendicular to the fault. As was mentioned, the copper rich rock types, breccias and C5 stockwork are closer to the fault, while the C1 and waste are more distant, but the rock types also vary in the other directions. For example, the breccias are wider towards the top. So we decided to use proportions which vary in all directions, hence to use a proportion matrix. This matrix has been built from 16 local East-West proportion curves, by kriging with a linear variogram.

With only four rock types, the rock-type rule is quite simple (Fig. 8.30). The first gaussian separates C1 from the breccias + C5 group on one side, waste on the other side, while the second separates Breccia and C5.

Once the rock-type rule has been chosen, the variograms can be fitted and the orebody can be simulated. The goal of this simulation was to propose a block model

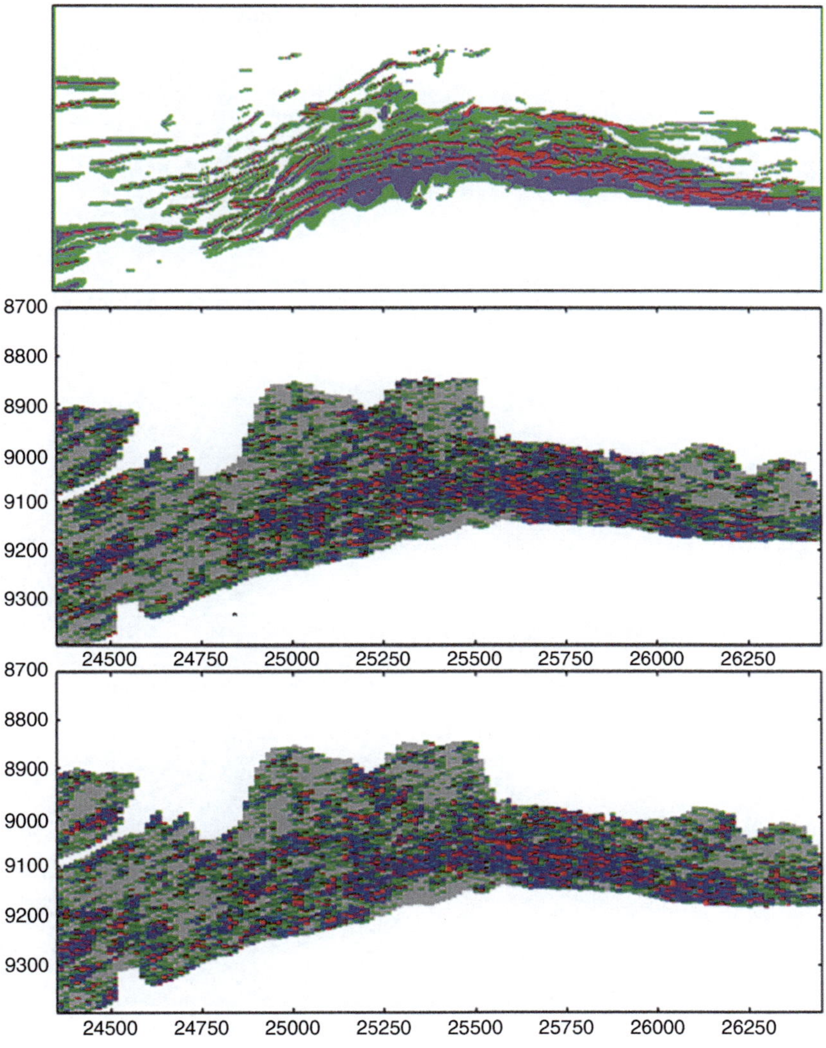

Fig. 8.31 Comparison between two simulations (center and below) and the block model given by the geologist (above). The waste rock (grey on the simulations) has not been represented on the geologist's block model. Reproduced from Riquelme et al. (2008) with permission

without manual smoothing done by the geologist. the new model compared favourably with the block model previously designed by the mine geologist.

Figure 8.31 compares the block model provided by the geologists (top) with two simulations. The waste rock has not been represented in the geologists model whereas it is grey in the simulations. An extended version of this case study can be found in Riquelme et al. (2008).

Chapter 9
Freeware

Introduction

In order to demonstrate the power and the flexibility of the truncated plurigaussian method, three free programs can be downloaded from the following Web site:

http://pluridemo.geosciences.mines-paristech.fr

PluriDemoSimu makes it possible to visualise simple plurigaussian simulations and to see how changes in the input parameters affect the resulting images.

PluriDemoVario illustrates the relationship between the indicator variogram and those of the underlying gaussian random functions (RFs for short) that are truncated to give the indicators.

PluriDemoSet allows the user to build a random set (two facies only) using the PluriGaussian Truncated method in a non-stationary framework.

For each program the web site also provides a set of graded exercises showing how to run the program and how the key parameters affect the simulated images or the indicator variograms, as the case may be.

Installation

The package is available for the following platforms:

- Windows (XP, Vista, Windows 7) 32/64 bits
- Linux x86 32/64 bits. Tests have been made under Ubuntu 9, Fedora 8, Redhat 4, Redhat 5.

Installation Instructions for Windows Users

1. Download the archive file named pluridemo_winnt.zip and save it into a temporary directory.

M. Armstrong et al., *Plurigaussian Simulations in Geosciences*,
DOI 10.1007/978-3-642-19607-2_9, © Springer-Verlag Berlin Heidelberg 2011

2. Unzip the archive file pluridemo_winnt.zip in the destination directory (for example in **C:\Programs File**).
3. Create the shortcuts for the three programs in the *Windows Start Menu* and on your desktop. Double click on the file named ***create_start_menu_links*** You will find this file in the directory where you installed the package (for example **C:\Programs File\PluriDemo**).
4. Start one of the programs by clicking on the corresponding icon created on your desktop or in the Windows Star Menu.

Installation Instructions for Linux Users

1. Download the archive file named pluridemo_linux.tgz and save it in a temporary directory (for example/***tmp***)
2. Open a terminal window
3. Choose an installation directory (for example/***usr/local***). Note that, depending on this choice, you may need the administrator privileges.
   cd /usr/local
   tar zxvf /tmp/pluridemo_linux.tgz.
4. Start one of the programs (for example: PluriDemoSimu) by typing the following command:
   /usr/local/PluriDemo/PluriDemoSim

Description of *PluriDemoSimu*

In this program, the user must select:

- The two underlying Gaussian random functions
- The rock type rule
- The proportions for each facies

Figure 9.1 presents a snapshot of the main panel of this program:

The Choice of the Model

Each underlying Gaussian random function (GRF) is chosen by selecting one of the 12 pre-programmed images. These images have been generated using three exponential models, three gaussian models, three cardinal sine models and three spherical models. In each case, the parameters are different, in particular the ranges and the anisotropy coefficients and directions (refer to Table 9.1 for the parameters).

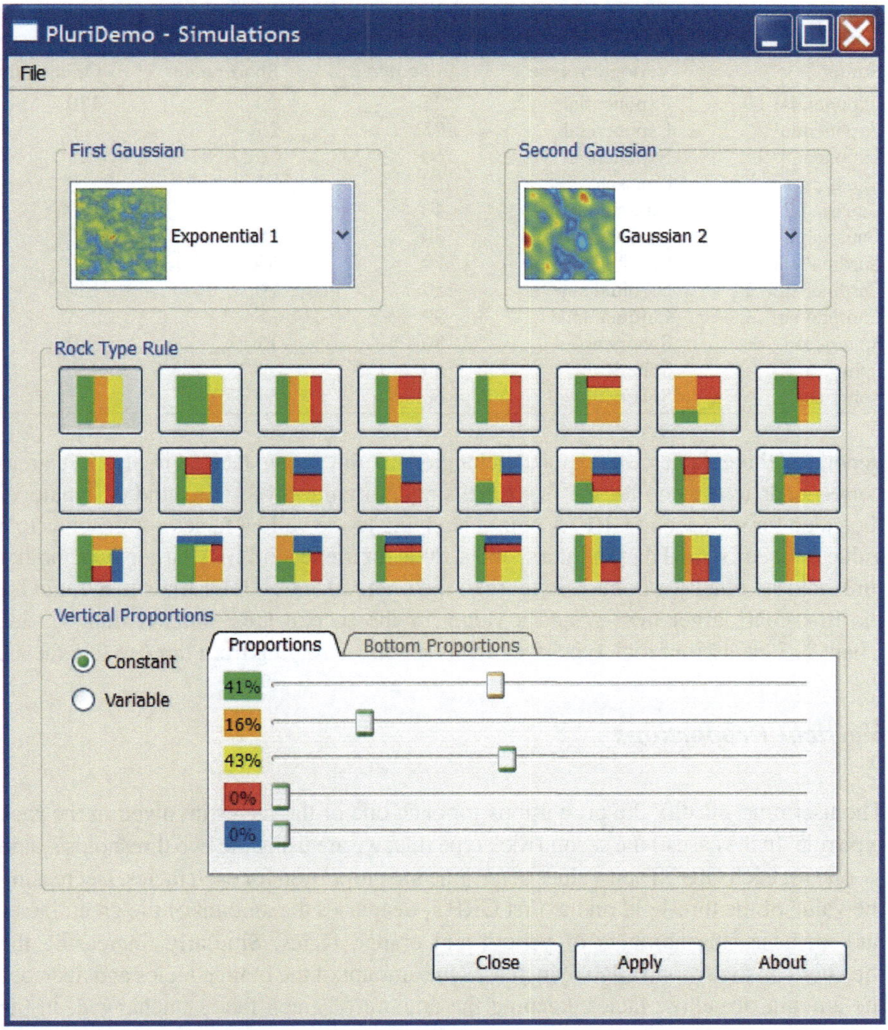

Fig. 9.1 PluriDemoSimu main window

The Rock Type Rule

The rock type rule defines the number of facies and the relationship between them.

A rock type rule is represented as a square icon which gives a symbolic representation of the thresholds separating facies. The horizontal axis representing the first GRF goes from $-\infty$ on the left to $+\infty$ on the right; and similarly for the second GRF along the vertical axis.

As an illustration, how do we interpret the second rock type rule of the first row? As the square is divided into three regions coloured orange, yellow and green, this rule

Table 9.1 Parameters of the variogram models used to generate the 12 images. (angles are measured counter-clockwise starting from east)

Name	Variogram type	Long range	Short range	Orientation
Exponential 1	Exponential	35	20	170°
Exponential 2	Exponential	105	25	40°
Exponential 3	Exponential	60	6	95°
Gaussian 1	Gaussian	50	7	0°
Gaussian 2	Gaussian	40	30	120°
Gaussian 3	Gaussian	40	18	60°
Cardinal sine 1	Cardinal sine	70	55	20°
Cardinal sine 2	Cardinal sine	150	20	140°
Cardinal sine 3	Cardinal sine	90	35	85°
Spherical 1	Spherical	80	15	10°
Spherical 2	Spherical	60	30	160°
Spherical 3	Spherical	60	50	35°

generates three facies. Any point belonging to the green facies (in the left area) corresponds to a value for the first GRF smaller than a given threshold s_1, whatever the value for the second GRF. A point belonging to the yellow facies corresponds to a value for the first GRF larger than s_1 and a value for the second GRF larger than another threshold s_2. Finally a point belonging to the orange facies corresponds to a value for the first GRF larger than s_1 and a value for the second GRF smaller than s_2. See Chap. 5. Twenty four rock type rules are available, involving from three to five facies.

Vertical Proportions

The user must specify the proportions for each one of the facies involved in the rock type rule. In the case of the second rock type rule, we are using the two thresholds s_1 and s_2, one for each GRF. Their values determine the proportion for each facies. Decreasing the value of the threshold on the first GRF s_1 decreases the amount of the green facies and increases the amounts of yellow and orange facies. Similarly, increasing the threshold on the second GRF s_2 increases the amount of the orange facies and decrease the amount of yellow facies, keeping the amount of green facies unchanged. In the default case the facies are programmed to have equal proportions (here 33%).

In the stationary case, these would be the same at all points in the area under study. The program also allows the user to check the impact of non-stationary proportions by defining the top and bottom proportion values. The proportions are then interpolated linearly from the top of the image to the bottom.

Simulation Outcome

When the parameters have been selected, the user produces the corresponding simulation outcome by pressing the Apply button.

The panel that appears on the output window is divided into four quadrants. The two images that were used in the construction are shown in the top two quadrants.

Note that the top right quadrant is left blank when the rock type rule indicates that the second GRF plays no role in the simulation. The resulting plurigaussian image appears in the bottom left quadrant. The bottom right square shows the rock type rule and the proportions that were used to generate it. The proportions are represented either as a pie diagram for stationary proportions or as a vertical proportion curve in the non-stationary case.

When the user performs several trials, the different simulation outcomes are produced in the same output window, subdivided into several tabs. When clicking on a given tab, the corresponding parameters are loaded back in the main window.

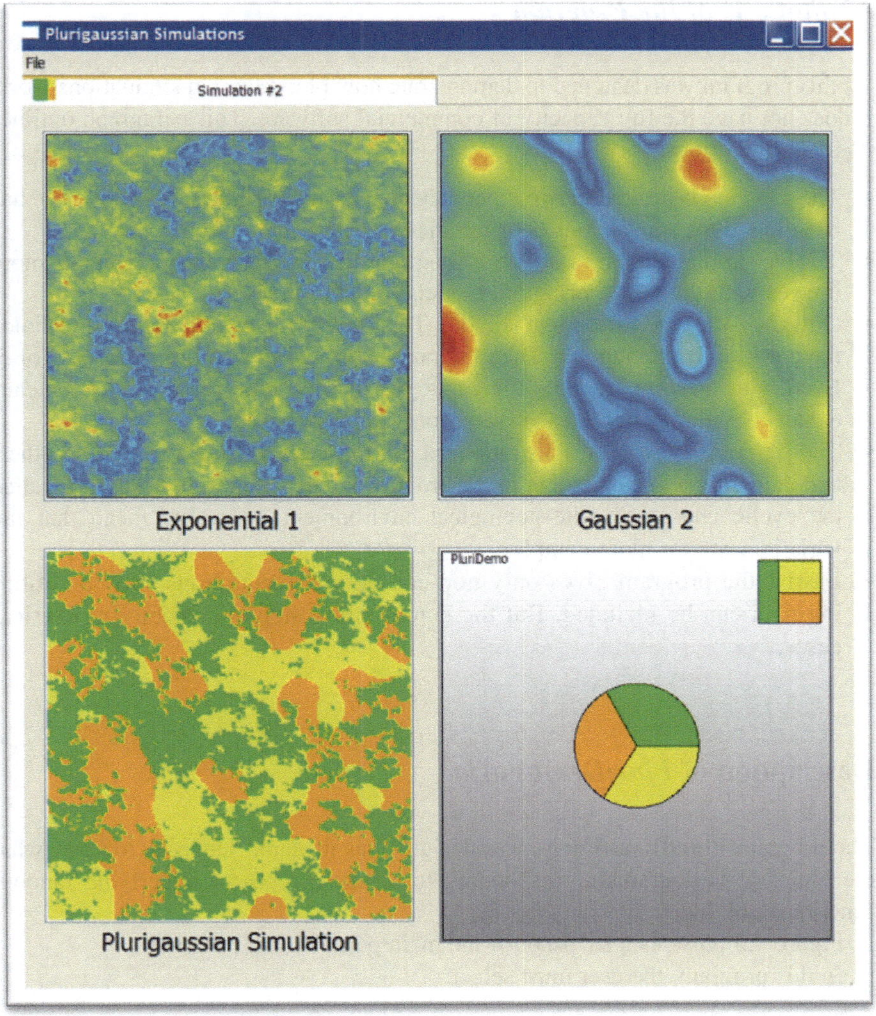

Fig. 9.2 PluriDemoSimu output window

Figure 9.2 represents the outcome of a plurigaussian simulation performed using the second rock type rule (involving three facies) when the first GRF corresponds to the model *Exponential 1* and the second GRF to the model *Gaussian 2*. All three facies have the same proportion.

The green foreground in the plurigaussian image corresponds to the lowest values of the first GRF. The contour lines around the green facies are broken up because its model is exponential. The contour separating the orange from the yellow is smooth because the corresponding model is gaussian.

Limitations of the Program

As this program was designed to demonstrate how plurigaussian simulations work, it does not have the full capacity of commercial software. This paragraph outlines its main limitations:

- The program is limited to two independent gaussian variables. No allowance has been made for correlations between the variables.
- Twelve simulated images were pre-programmed. The user cannot generate more images with the same variogram models or with others.
- Only 24 rock type rules are available. To simplify the calculation of thresholds the partition is rectangular. Then the number of facies was limited to 3, 4 or 5. Even with these restrictions the catalogue of rock type rules is by no means exhaustive. For example each colour only appears once.
- The program allows for non-stationarity in the vertical direction (but no other) and then only a linear change in the proportions from top to bottom. In practice the cyclic changes in the geological environment over time mean that the variations are far more complex.
- Lastly, the program gives only non-conditional simulations. No conditioning data can be included. But the conditioning step is possible in practical cases.

Description of *PluriDemoVario*

The program **PluriDemoVario**, was designed, as its name suggests, to show what the indicator variograms corresponding to different types of underlying random functions look like.

Figure 9.3 presents a snapshot of the main panel of this program:

In this program, the user must select:

- The model for the two underlying gaussian random functions
- The rock type rule

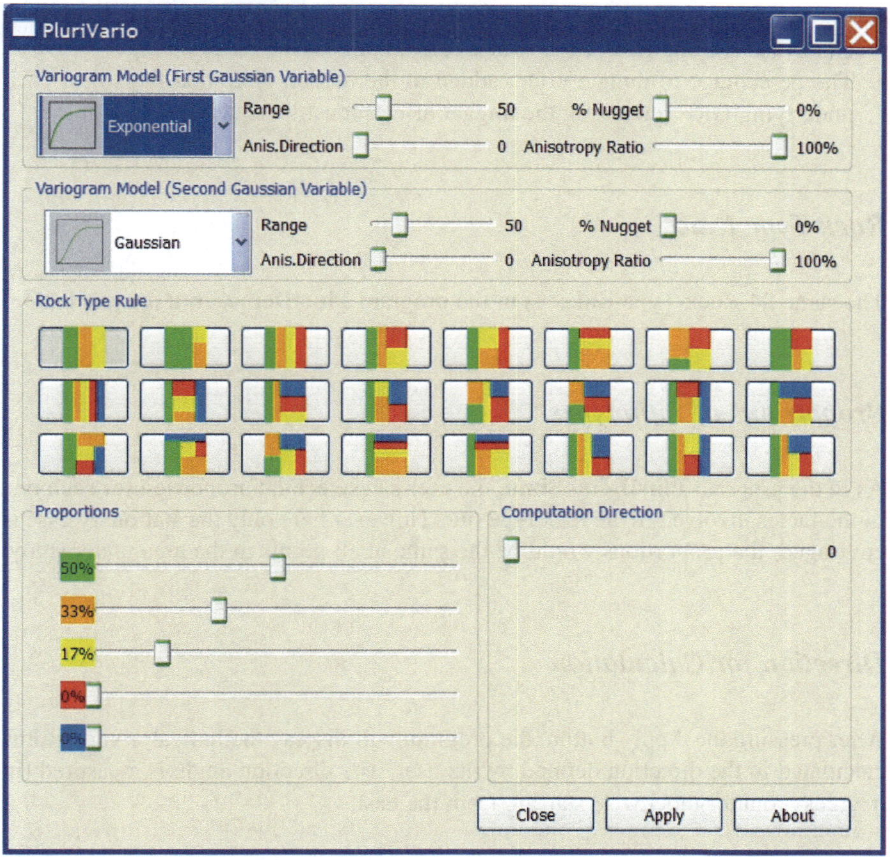

Fig. 9.3 PluriDemoVario main window

- The proportions for each facies
- The direction in which the variogram of indicators will be calculated.

Choice of the Model

The user must specify the type of variogram for each underlying GRF. Unlike in **PluriDemoSimu**, the user can choose:

- The variogram type selected among the following possibilities: exponential, gaussian, cardinal sine or spherical
- The (generic) range
- The anisotropy ratio which gives the ratio between the smallest and the longest ranges

- The anisotropy direction (an angle between 0° and 360° measured counter-clockwise starting from east) which corresponds to the longest range
- The percentage of nugget effect added to the model. Note that the sill of each underlying GRF (including the nugget effect) must be equal to 1.

Rock Type Rule

The same 24 Rock Type Rules as in the program **PluriDemoSimu** are available.

Proportions of Lithotypes

As in the program **PluriDemoSimu**, the user can select the proportion for each one of the facies involved in the rock type rule. However here only the stationary case is envisaged: the proportions would be the same at all points in the area under study.

Direction for Calculation

After pressing the Apply button, the program will display the indicator variograms calculated in the direction defined by the user. The direction angle is measured (in degrees) counter-clockwise starting from the east.

Indicator Variograms

When the parameters have been selected, the user produces the corresponding indicator variograms by pressing the Apply button.

The output window is divided into four quadrants. The top two images represent the model selected for each underlying GRF. Note that the top right quadrant is left blank when the rock type rule indicates that the second GRF plays no role. The bottom left quadrant represents the 2D anisotropy diagram showing the distance at which the variogram of each indicator reaches 95% of its sill value as a function of direction. The bottom right quadrant represents the variograms of the indicators calculated in the direction requested by the user. It also presents a set of icons which allow the user to select:

- **S**: simple variograms of indicators
- **SN**: simple variograms of indicators after their sills have been normalized to 1
- **X**: cross-variograms of indicators between the reference and all the other facies. The reference facies is selected using the corresponding coloured icon

- **XN**: cross-variogram of indicators between the reference and all the other facies, after their sills have been normalized
- **F**: the function measuring the edge effect of a reference against the other facies.

When the user performs several trials, the different indicator variograms are produced in the same output window, subdivided into several tabs. By clicking on a given tab, the corresponding parameters are loaded back in the main window.

Figure 9.4 represents the outcome of the program **PluriDemoVario** performed using:

- An exponential model with a range of 50 and anisotropy with a ratio of 51% and a direction of 30° for the first GRF
- A gaussian model with a range of 50 and anisotropy with a ratio of 70% and a direction of 61° for the second GRF

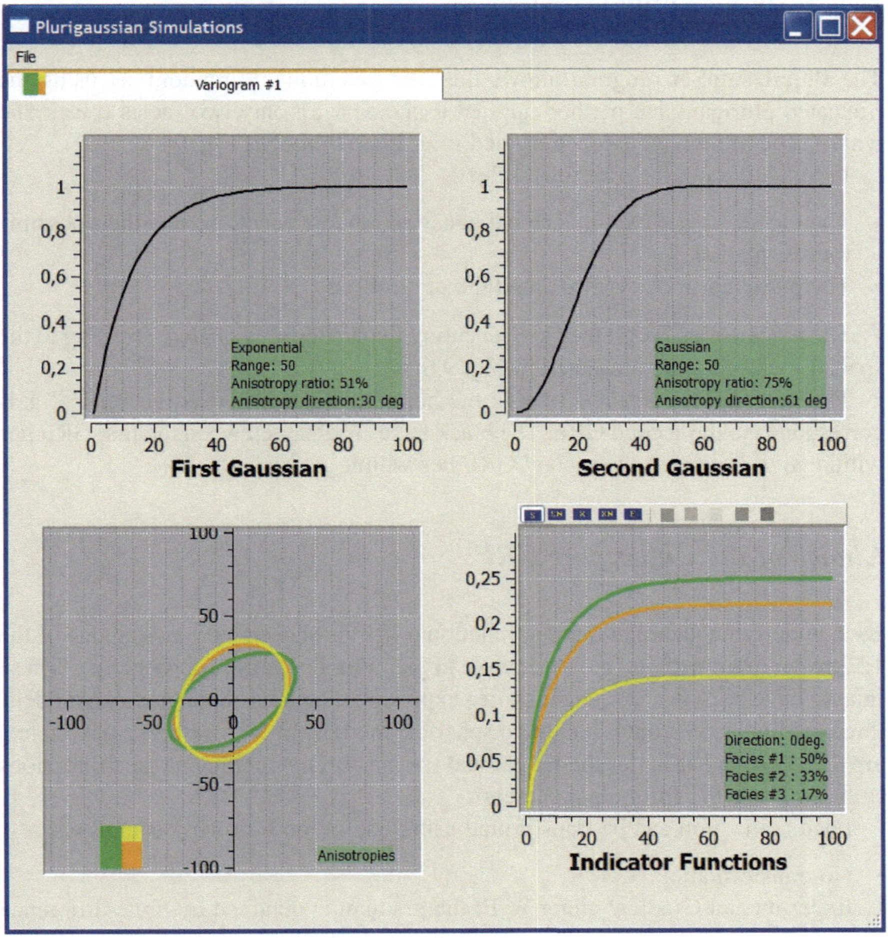

Fig. 9.4 PluriDemoVario output window

The second rock type rule involving three facies with the following proportions: proportions for the facies are 50% green, 33% orange and 17% yellow.

The 2D anisotropy diagram shows the green ellipse with a main direction oriented 30° and a ratio of the smallest to the longest ranges equal to 0.5 (51%). This reflects that the green facies only depends upon the model characteristics of the first GRF. The curves for the other facies are not precisely ellipses as they combine the model characteristics of the two GRFs, and therefore their anisotropy directions which do not coincide (30° for the first GRF and 61° for the second GRF).

The variogram of indicators are all different (this can be checked using the normalized version). The one corresponding to the green facies is the translation of the exponential variogram of the underlying first GRF to the indicator, whereas the other two indicator variograms involve the models of both underlying GRFs.

Description of PluriDemoSet

The **PluriDemoSet** program allows the user to simulate a random set using the truncated plurigaussian method applied to a case with only two facies (black and white). The main window is presented in of **PluriDemoSet** Fig. 9.5.

In this program, the user must select:

- The model for the two underlying gaussian random functions and their possible transformation
- The particular rock type rule and the proportions

The black facies is obtained by truncating two underlying GRFs according to the special rock type rule presented in Fig. 9.6:

The first GRF is displayed along the horizontal axis whereas the vertical axis corresponds to the second GRF. The black facies is obtained when the first GRF lies within $[s_1; s_2]$ or when the second GRF lies within $[t_1; t_2]$.

Choice of the Underlying GRF

Each underlying gaussian random function (GRF) is chosen by selecting one of the 12 pre-programmed images (same as in the **PluriDemoSimu** program). These images have been generated using three exponential models, three gaussian models, three cardinal sine models and three spherical models. In each case, the parameters are different, in particular the ranges and the anisotropy coefficients and directions (refer to Table 9.1 for the parameters).

Each realization can be transformed using one of the following operations:

- No transformation
- Experimental Gradient along W-E: the gradient is obtained by finite difference comparing consecutive grid nodes along the W-E axis
- Experimental Gradient along N-S

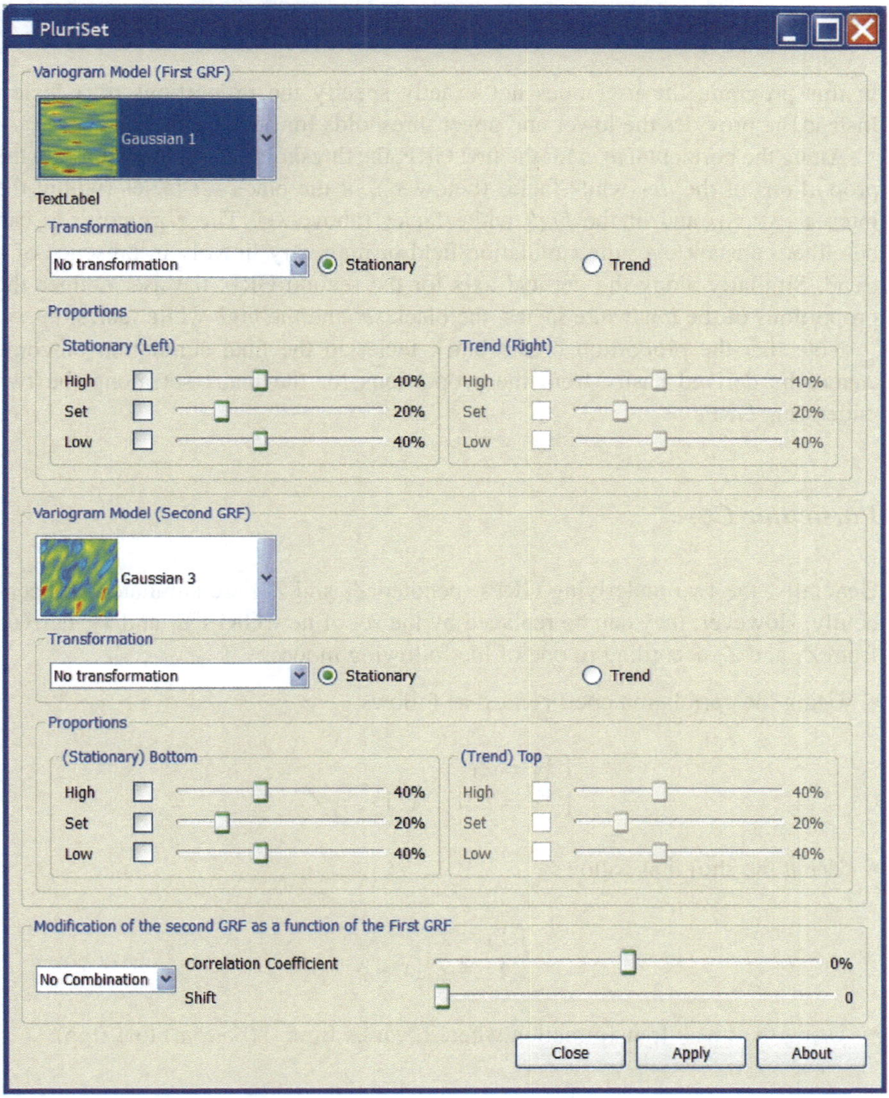

Fig. 9.5 PluriDemoSet main window

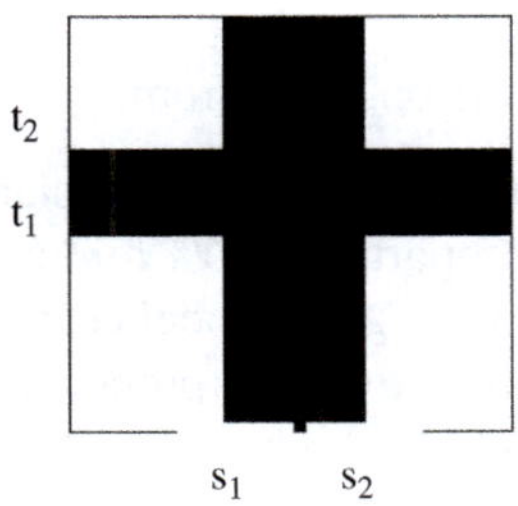

Fig. 9.6 Rock Type Rule

The Proportions

In this program, the user does not exactly specify the proportions of a facies. Instead, he provides the lower and upper thresholds for both GRFs.

Along the horizontal axis for the first GRF, the thresholds are defined through the proportions of the *low* white facies (below s_1), of the black *set* facies (within the interval $[s_1; s_2]$) and of the *high* white facies (above s_2). These proportions can be either constant over the simulation field or may vary linearly in presence of a trend. Similarly along the vertical axis for the second GRF, the user defines the proportions of the *low* white facies, the black *set* and the *high* white facies.

Note that the proportion of the black facies in the final simulation outcome cannot be derived easily from the proportions for the black set along the two underlying GRFs.

Particular Case

Generally, the two underlying GRFs (denoted Z_1 and Z_2) are simulated independently. However, they can be replaced by the set of new GRFs Y_1 and Y_2, derived from Z_1 and Z_2 according to one of the following modes:

- Using the correlation coefficient ρ as follows:

$$\begin{cases} Y_1 = Z_1 \\ Y_2 = \rho Z_1 + \sqrt{1-\rho^2} Z_2 \end{cases}$$

- Using the shift δ as follows:

$$\begin{cases} Y_1 = Z_1 \\ Y_2 = Z_1(x+\delta) \end{cases}$$

- Using the linear transformation where α varies from 0 (bottom) to 1 (top):

$$\begin{cases} Y_1 = Z_1 \\ Y_2 = \alpha Z_1(x) + (1-\alpha) Z_2(x) \end{cases}$$

In Fig. 9.7, the random set is obtained from:

- The first GRF generated using the model Gaussian 1 and showing a strong W-E anisotropy, with constant proportions: 40% for low, 20% for set and 40% for high
- The second GRF generated using the model Gaussian 3 and showing a moderate anisotropy along the first bisector, with proportions varying from bottom (40% for low, 20% for set and 40% for high) to top (65% for low, 31% for set and 3% for high)

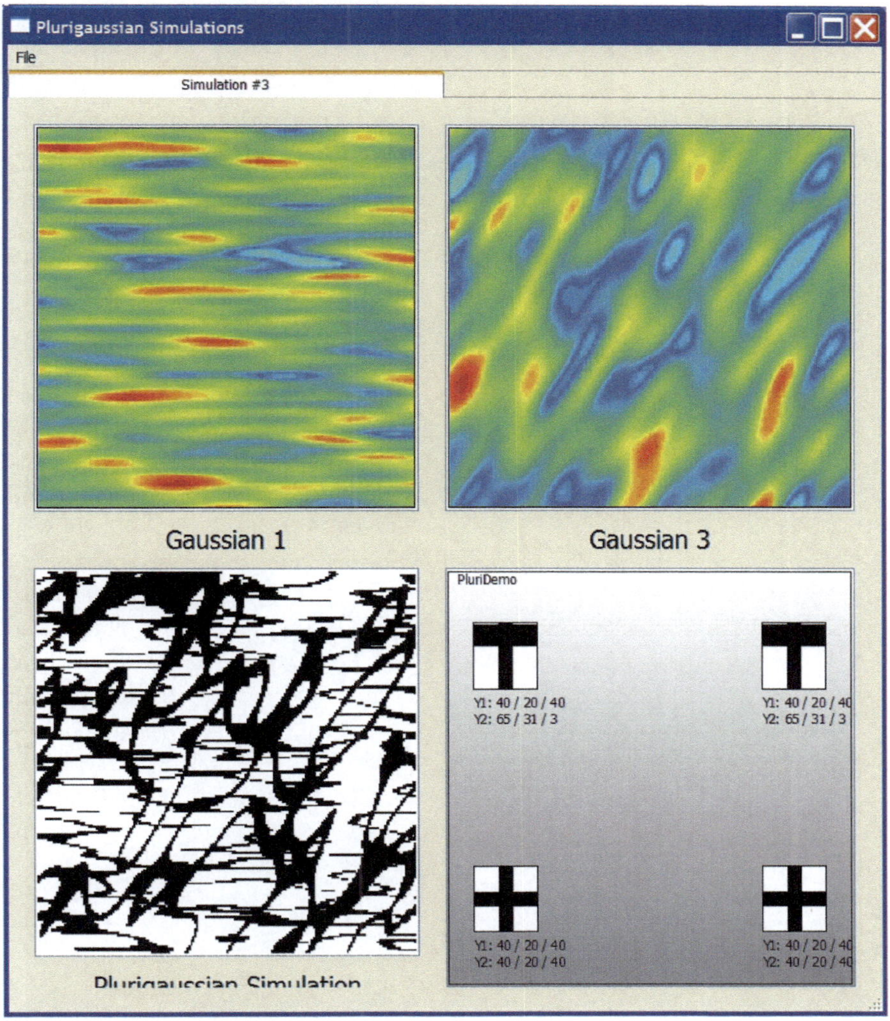

Fig. 9.7 PluriDemoSet output window

The output window is subdivided into four quadrants:

- The top quadrants show the underlying GRFs used for the simulation
- The bottom right quadrant shows the rock type rule at the four corners of the field, represented with the corresponding proportions
- The bottom left quadrant represents the simulated outcome

We can check that the black facies is a mixture of thin stripes oriented along the W-E axis and thicker meanders oriented along the first bisector. These meanders become even thicker in the topmost part of the simulation (due to the trend in the proportions).

References

Aanonsen SI, Naevdal G, Oliver DS, Reynolds AC, Vallès B (2009) The Ensemble Kalman filter in reservoir engineering – review, SPE 117274. SPE J 14(3):393–412

Agbalaka CC, Oliver DS (2008) Application of the EnKF and localization to automatic history matching of facies distribution and production data. Math Geosci 40:353–374, 2008

Agbalaka CC, Oliver DS (2009) Automatic history matching of production and facies data with non-stationary proportions using EnKF. SPE 118916, presented at 2009 Reservoir simulation symposium held in Woodlands, Texas, 2–4 Feb 2009

Albertao GA, Grell AP, Badolato D, dos Santos LR (2005) 3D geological modeling in a turbidite system with complex stratigraphic-structural framework – an example from Campos Basin, Brazil. SPE 95612, presented at the 2005 SPE Annual technical conference & exhibition held in Dallas, Texas, 9–12 Oct, 2005

Amit Y, Grenader U (1991) Comparing sweep strategies for Stochastic Relaxation. J Multivariate Anal 37:197–222

Armstrong M (1992) Positive definiteness is not enough. Math Geol 1:135–143

Armstrong M, Galli A (1999) Derivative based plurigaussian simulations. In: Lippard SJ, Naess A, Sinding-Larsen R (eds) IAMG 99, Trondheim, pp 591–596

Azpiritxaga I, Hernandez E, Correa A (1997) Uso del modelaje geostadistico gaussiano truncado en la definicion de secuencias estratigraficas en yaciemento C6 del area VLA-8, bloque I, Lago de Maracaibo. In: Memorias del I Congreso Latinoamericano de Sedimentologia, Soc Venezolana de Geol. Tomo 1, 59–65, Nov 1997

Babish G (2006) Geostatistics without tears: a practical guide to surface interpolation, Geostatistics, Variograms and Kriging. Environment Canada, Gatineau, Quebec, 117 pp

Barone P, Frigessi A (1990) Improving stochastic relaxation for gaussian random fields. Probab Eng Inform Sci 4:369–389

Besag J (1974) Spatial interaction and the statistical analysis of lattice systems (with discussion). J Royal Statist Soc B 36:192–236

Betzhold J, Roth C (2000) Characterising the mineralogical variability of a Chilean copper deposit using plurigaussian simulations, J South African IMM, March/April, pp111–120, available from http://www.saimm.co.za/Journal/v100n02p111.pdf

Beucher H, Galli A (1994) La géostatistique dans le domaine pétrolier. In: Proceedings of the 4th Tunisian petroleum exploration conference, Tunis pp 379–393

Beucher H, Galli A, Le Loc'h G, Ravenne C (1993) Including a regional trend in reservoir modelling using the truncated Gaussian method. In: Soares A (ed) Geostat Troia '92, vol 1. Kluwer, Dordrecht, pp 555–566

Beucher H, Geffroy F, Doligez B (1997) Introducing more geology in reservoir stochastic modelling using truncated Gaussian approach. In: Memorias del I Congreso Latinoamericano de Sedimentologia. Soc Venezolana de Geol. Tomo 1, 89–95 Nov, 1997

M. Armstrong et al., *Plurigaussian Simulations in Geosciences*,
DOI 10.1007/978-3-642-19607-2, © Springer-Verlag Berlin Heidelberg 2011

Beucher H, Fournier F, Doligez et al (1999) Using 3D seismic-derived information in lithofacies simulations: a case study. Proc of the 1999 SPE Annual Technical Conference and Exhibition, Houston, Tx., 3–6 Oct. 1999, Houston, Tx. pp. 581–192. SPE 56736

Beucher H, Doligez B, Yarus JM (2006) Modeling complex reservoirs with multiple conditional techniques: a practical approach to reservoir characterization, *AAPG computer application in Geology* 5, Stochastic modeling and geostatistics: principles, methods and case studies vol. II 289

Bianco A, Cominelli A, Dovera L, Naevdal G, Vallès B (2007) History mathching and production forecasting uncertainty by means of the Ensemble Kalman Filter: a real field application. SPE paper 107161, presented at the EAGE/EUROPEC conference & exhibition, London, 11–14 June, doi: 10.2118/107161-MS

Biver P, Haas A, Bacquet C (2002) Uncertainties in facies proportion estimation II: application to geostatistical simulation of facies and assessment of volumetric uncertainties. Math Geol 34(6):703–714

Cacas MC, Daniel JM, Letouzey J (2001) Nested geological modelling of naturally fractured reservoirs. Petrol Geosci 7:S43–S52

Carrasco P, Ibarra F, Rojas R, Le Loc'h G, Seguret S (2007) *Application of the Truncated Gaussian Simulation Method to a Porphyry Copper Deposit*. In: Magri Ed., APCOM 2007, 33rd International Symposium on Application of Computers and Operations research in the Mineral industry, pp 31–39

Chatelin F (1983) Spectral approximation of linear operators. Academic, New York

Chautru JM, Ravenne C, Lemouzy P, Galli A (1992) Heresim 3D: integrating data and disciplinaries for reservoir characterization. In: Annales du 4eme EC Symposium, Berlin, 11 p

Chen Y, Oliver DS (2010) Parametrization techniques to improve mass conservation and data assimilation for Ensemble Kalman Filter. SPE 133560, Presented at the SPE Western Regional Meeting held in Annaheim, 27–29 May, 2010

Cherubini C, Giasi CI, Musci F, Pastore N (2009a) Application of truncated plurigaussian method for the reactive transport modeling of a contaminated aquifer. In: Recent advances in water resources, hydraulics & hidrology – mathematics and computers in science and engineering. WSEAS Press. ISBN: 978-960-474-057-4 ISSN: 1790–2769. (ISI Book)

Cherubini C, Giasi CI, Musci F, Pastore N (2009b) Checking simulations of a geolithological model obtained by means of nested truncated bigaussian method. In: Int J Math Model Meth Appl Sci 3(2):152–161

Chilès JP, Delfiner P (1999) Geostatistics: Modeling spatial Uncertainty, John Wiley and Son, New York, 695pp

Cosentino L, Coury Y, Daniel JM, Manceau E, Ravenne E, Van Lingen P, Cole J, Sengul M (2001) Integrated study of a fractured Middle East reservoir with stratiform super-K intervals – Part 2: Upscaling & dual media simulation. SPE 68184, presented at the 2001 SPE Middle East Oil Show, held in Bahrain, 17–20 Mar 2001

Chopin N (2011) Fast simulation of truncated gaussian distributions. Stats & Comput 21(2): 275_288 doi: 10.1007/s11222-009-9168-1

Cowles MK, Carlin BP (1996) Markov Chain Monte Carlo convergence diagnostics: a comparative review. J Am Stat Assoc 91:883–904

Cozzi M, Roncarolo F, Balossino P, Bersezio R, Felletti F, Rossi M (2002) Combining facies analysis, log interpretation and geostatistical simulation to characterise heterogeneity of a turbiditic reservoir. In: IAMG'02 Proceeding of the 8th Annual Conference of the International Association for Mathematical Geology, Berlino (eds. U. Bayer, H. Burger, W. Skala)

Da Veiga S, Le Ravalec M (2010) Rebuilding existing geological models. SPE 130976 presented at the SPE EUROPEC/EAGE annual conference & exhibition, held in Barcelona, Spain, 14–17 June 2010

de Fouquet C, Beucher H, Galli A, Ravenne C (1989) Conditional simulation of random sets: application to an argilaceous sandstone reservoir. In: Armstrong M (ed) Geostatistics, vol 2. Kluwer, Dordrecht, pp 517–530 de Galard J-H, Zoormand GH, Ghanizadeh M, Daltaban S Camus D (2005) A case study on redevelopment of a giant highly fractured carbonate reservoir in Iran based on integrated reservoir characterization and 3D modeling studies. SPE 93760 presented at the 14th SPE Middle East oil & gas show & conference, held in Bahrain, 12–15 Mar 2005

Deraisme J, Field M (2006) Geostatistical simulations of kimberlite orebodies: application to sampling optimisation. Proceedings of the 6th international mining geology conference, Darwin, NT, Australia, 21–23 Aug 2006, pp 193–203

Deutch C, Journel AG (1992) GSLIB Geostatistical software library and user's guide. Oxford University Press, New York

Deutsch C (2006) A sequential indicator simulation program for categorical variables with point & block data: BlockSIS. Comput Geosci 32:1669–1681

Devroye L (1986) Non-Uniform Random Variate Generation. Springer, New York

Doligez B, Ravenne C, Galli A, Lemouzy P (1992) Une methodologie pour une etude integree des réservoirs: des données de puits aux simulations d'écoulement en utilisant des outils géostatistiques. Pét Tech 372:43–48

Doligez B, Beucher H, Fonnesu F, Tebaldi E, Baruffini L (1994a) Reservoir modeling: the reference model: Cajigar. In: Helbig K (ed) Modeling the earth for oil exploration. Pergamon, Brussels, pp 81–91

Doligez B, Beucher H, Fonnesu F, Tebaldi E, Baruffini L (1994b) Reservoir modeling: limited data sets: Cajigar. In: Helbig K (ed) Modeling the earth for oil exploration. Pergamon, Brussels, pp 108–110

Doligez B, Beucher H, Geffroy F, Eschard R (1999a) Integrated reservoir characterisation: improvement in heterogeneous stochastic reservoir modeling by integration of additional external constraints. In: Schatzinger R, Jordans J (eds) Reservoir characterization – recent advances, vol 71, AAPG Memoir., pp 333–342

Doligez B, Granjeon D, Joseph P, Eschard R, Beucher H (1999b) How can stratigraphic information help constrain geostatistical reservoir simulations? In: Numerical experiments in stratigraphic and sedimentologic computer simulations. SPEM Special Publications N° 62, pp 239–244

Doligez B, Eschard R, Joseph P, Beucher H (1999c) Une méthodologie intégrée pour l'estimation des volumes en place dans les réservoirs et la quantification de leurs incertitudes. Pét Tech 422:50–54

Dowd PA (1997) Structural controls in the geostatistical simulation of mineral deposits. In: Baafi E et al (eds) Geostatistics Wollongong '96, vol 2. Kluwer, Dordrecht, pp 647–657

Dowd PA, Pardo-Iguzquiza E, Xu C (2003) Plurigau: a computer program for simulating spatial facies using the truncated plurigaussian method. Comput Geosci 29:123–141

Dowd PA, Xu C, Mardi KV, Fowler RJ (2007) A comparison of methods for stochastic simulation of rock fractures. Math Geol 39:697–714, 2007

Emery X (2002) Conditional simulation of nongaussian random functions. Math Geol 34:79–100

Emery X (2005) Conditional simulation of random fields with bivariate gamma isofactorial distributions. Math Geol 37:419–445

Emery X (2007a) Using the Gibbs sampler for conditional simulation of Gaussian-based random fields. Comput Geosci 33:522–537

Emery X (2007b) Simulation of geological domains using the plurigaussian model: new developments and computer programs. Comput Geosci 33:1189–1201

Emery X (2008) A turning bands program for conditional co-simulation of cross-correlated Gaussian random. Comput Geosci 34:1850–1862

Emery X (2010) Iterative algorithms for fitting a linear model of coregionaliszation. Comput Geosci 39(9):1150–1160

Emery X, Gonzalez KE (2007a) Probabilistic modelling of mineralogical domains and its application to resources evaluation. J South African IMM 107(12):803–809

Emery X, Gonzalez KE (2007b) Incorporating the uncertainty in geological boundaries into mineral resources evaluation. J Geol Soc India 69(1):29–38

Emery X, Lantuejoul C (2006) TBSIM: a computer program for conditional simulation of three-dimensional Gaussian random fields via the turning bands method. Comput Geosci 32:1615–1628

Emery X, Silva DA (2009) Conditional co-simulation of continuous and categorical variables for geostatistical applications. Comput Geosci 35:1234–1246

Eschard R, Doligez B, Rahon D, Ravenne C, Le Loc'h G (1991) A new approach for reservoirs description and simulation using geostatistical methods. In: Advances in reservoir technology, characterization, modelling and management, Edinburgh, 13p

Eschard R, Desaubliaux G, Houel P, Beucher H (1992) Geometry of an incised valley-fill complex (Southbay, Middle Jurassic, Yorkshire). In: 7th IFP Research conference in exploration production, subsurface characterization from outcrop observations, Scarborough, Yorkshire

Eschard R, Lemouzy P, Bacchiana C, Desaubliaux G, Parpant J, Smart B (1998) Combining sequence stratigraphy, geostatistical simulations, and production data for modeling a fluvial reservoir in the Chaunoy field (Triassic, France). AAPG Bulletin, Vol. 82, No. 4, April 1998, pp 545–658

Eschard R, Lemouzy P, Bacchiana C, Désaubliaux G, Parpant J, Smart B (1999) Combining sequence stratigraphy, geostatistical simulations and production data for modeling a fluvial reservoir in the Chaunoy field (Triassic, France). AAPG Bull 82:545–568

Eschard R, Doligez B, Beucher H (2002) Using quantitative outcrop databases as a guide for geological reservoir modelling. In: Armstrong M et al (eds) Geostatistics Rio 2000. Kluwer, Dordrechts, Holland, pp 7–18

Evensen G, Hove J, Meisinger HC, Reiso E, Seim KS, Espelid O (2007) Using EnKF for assisted history matching of a North Sea reservoir model. SPE 106184, presented at the 2007 SPE Reservoir Simulation symposium, held in Woodlands, Texas, 26–28 Feb, 2007

Felletti F (2004) Statistical modelling and validation of correlation in turbidites: an example from the Tertiary Piedmont Basin (Castagnola Fm., North Italy). Marine Petroleum Geology, Vol. 21 pp 23–39

Fontaine L, Beucher H (2006) Simulation of the Muyumkum uranium roll front deposit by using truncated plurigaussian method. In: Proceedings of the 6th international mining geology conference, Darwin, NT Australia, 21–23 Aug 2006

Fournier F, Derain JF (1997) A statistical methodology for deriving reservoir properties from seismic data. Geophysics 60(5):1437–1450

Freulon X (1992) Conditionnement du modèle gaussien par des inégalités ou des randomisées. Thèse de Docteur en Géostatistique, ENSMP. 168p

Freulon X (1994) Conditional simulation of a Gaussian random vector with nonlinear and/or noisy observations. In: Armstrong M et al (eds) Geostatistical simulations. Kluwer, Dordrecht, pp 57–71

Freulon X, de Fouquet C (1993) Conditioning a Gaussian model with inequalities. In: Soares A (ed) Geostat Troia '92, vol 1. Kluwer, Dordrecht, pp 201–212

Freulon X, de Fouquet C, Rivoirard J (1990) Simulation of the geometry and grades of a uranium deposit using a geological variable. In: Proceedings of the XXII international symposium APCOM. Vol 2, pp 649–659

Galli A, Beucher H (1997) Stochastic models for reservoir characterization: a user-friendly review. In: Fifth Latin American and Caribbean petroleum engineering conference and exhibition, Rio de Janeiro, Brazil, 30 Aug–3 Sept 1997. SPE 38999, 11 p

Galli A, Guérillot D, Ravenne C (1990) Combining geology, geostatistics and multiphase fluid flow for 3D reservoir In: Guérillot D et al (eds) 2nd European conference on the mathematics of oil recovery. Technip, Paris, pp 11–19

Galli A, Beucher H, Le Loc'h G, Doligez B (1994) The pros and cons of the truncated gaussian method. In: Armstrong M et al (eds) Geostatistical simulations. Kluwer, Dordrecht, pp 217–233

Galli A, Ravenne C, Richard V, Guérillot D (1995) Constraining detailed geostatistical reservoir models with seismic and dynamic data. In: Seamless seismic to simulation and back. Proceedings of EAPG/SPE Workshop, Glasgow, 6 p

Galli A, Gao H (2001) Rate of convergence of the Gibbs Sampler in the Gaussian Case. Math Geol 33(6):653–678

Galli A, Le Loc'h G, Geffroy F, Eschard R (2006) An Application of the Truncated Pluri-gaussian Method for Modeling Geology. AAPG computer application for geology, pp 109–122

Gao G, Zafari M, Reynolds AC (2006) Quantifying uncertainties for the PUNQ-S3 problem in a Bayesian setting with RML and EnKF. SPE J 11(4):506–515, SPE-93324-PA, doi 10.2118/93324-PA

Geman S, Geman D (1984) Stochastic relaxation, Gibbs distribution and the bayesian restoration of images. IEEE Trans Pattern Anal Mach Intell 6:721–741

Gervais V, Roggero F, Le Ravalec M (2009) Adaptative local parametrization of facies proportions for the history matching of production and time lag saturation data. IFP Working Paper, 10 pp

Geweke J (1991) Efficient simulation from the multivariate normal and Student-t distributions subject to linear constraints and the evaluation of constraint probabilities. Comput Sci Stat 23:571–578

Gikhman IJ, Skorokhod AV (1969) Introduction to the theory of random processes. Dover, New York

Gossa H, Gaaya MH, Yahmadi B, Volpi B, Doligez B, Hu LY, Galli A (1993) A geostatistical approach aimed at reservoir characterization: a case study. In: 2èmes Journées Tunisiennes de Géologie Appliquée, Sfax, Tunisie, 10 p

Grammer GM, Eberli GB, Van Buchem FSP, Stevenson GM, Homewood P (1996) Application of high-resolution sequence stratigraphy to evaluate lateral variability in outcrop and subsurface- Desert Creek and Ismay intervals, Paradox Basin. In: Longman MW, Sonnenfeld MD (eds) Paleozoic systems of the Rocky Mountain region, Rocky Mountain Section, SEPM, pp 235–266

Grammer GM, Eberli GP, Van Buchem FSP, Stevenson GM, Homewood PW (2000) Application of high resolution sequence stratigraphy in developing an exploration and production strategy for a mixed carbonate/siliciclastic system (Carboniferous), Paradox Basin, USA. In: Homewood PW, Eberli GP (eds) Genetic Stratigraphy on the Exploration and Production Scales: Case Studies from the Pennsylvanian of the Paradox Basin and the Upper Devonian of Alberta, Elf Aquitaine Memoir 24, pp 29–69

Grötsch J, Mercadier C (1999) Integrated 3-D Reservoir modeling based on 3-D seismic: the Tertiary Malampaya and Camago buidups, offshore palawan, Philippines. AAPG Bulletin 83(11):1703–1728

Gu Y, Oliver DS (2005) History matching of the PUNQ-S3 reservoir model using the Ensemble Kalman Filter. SPE J 10(2):217–224, SPE-89942-PA, doi: 10.2118/89942-PA

Gu Y, Oliver DS (2007) An iterative Ensemble Kalman filter for multiphase fluid flow data assimilation, SPE 108438. SPE journal Dec 2007:438–446

Guérillot D, Rudkiewicz JL, Ravenne C, Renard D, Galli A (1989) An integrated model for computer aided reservoir description: from outcrop study to fluid flow simulations. In: Proceedings of the 5th European symposium on improved oil recovery, Budapest. pp 651–660

Guérillot D, Beucher H, Galli A, Lemouzy P, Morelon I (1990) Multiphase fluid flow for 3D geostatistical reservoir description. In: Proceedings of the OMMBKE 90, 21st Petroleum itinerary conference and exhibition, Siofok, Hungary. pp 252–261

Guérillot D, Lemouzy P, Galli A, Ravenne C (1990) 3D fluid flow behaviour in porous media characterized by geostatistical methods. In: SPE Latin American petroleum engineering conference and exhibition, Rio de Janeiro. SPE 21081. 8p

Gunning J, Glinsky M, White C (2007) Delivery Massager: a tool for propagating seismic inversion information to reservoir models. Comput Geosci 33:630–648

Haugen V, Natvik L-J, Evensen G, Berg A, Flornes K, Naevdal G (2006) History matching using the Ensemble Kalman filter on a North Sea field case. SPE 102430, Presented at the 2006 SPE annual tecnical conference & exhibition, held in San Antonio, Texas, 24–27 Sept and published in the SPE J, Dec 2008, pp 382–391

Haugen V, Naevdal G, Natvik L-J, Evensen G, Berg AM, Flornes KM (2008) History matching using the Ensemble Kalman Filter on a North Sea field case. SPE J 13(4):382–391, SPE-102430-PA, doi: 10.2118/102430-PA

Hendriks Franssen HJ, Alcolea A, Riva M, Bakr M, van der Wiel N, Stauffer F, Guadagnini A (2009) Comparison of seven methods for inverse modelling of groundwater Application to the characterization of well catchments. Adv Water Resour 32:851–872

Hu L-Y (2000) Gradual deformation and iterative calibration of Gaussian-related stochastic models. Math Geol 32(1):87–108

Hu LY, Le Ravalec M, Blanc G, roggero F, Noetinger B, Haas A Corre B (1999) Reducing uncertainties in production forecasting by conditioning to dynamic data. SPE 56703, presented at the 1999 SPE annual technical conference & exhibition, held in Houston, 3–6 Oct 1999

Jacod J, Joathon P (1971) Use of random genetic models in the study of sedimentary processes. Math Geol 3(3):265–279

Johann P, Fournier F, Souza O, Eschard R, Beucher H (1996) 3D stochastic modeling constrained by well and seismic data on a turbidite field. SPE 36501, SPE annual technical conference and exhibition, Denver, CO, 6–9 October

Kerrou J, Renard P, Henricks-Franssen H-J, Lunati I (2008) Issues in characterizing heterogzneity & connectivity in non-Gaussian media. Adv Water Resour 31:147–159

Kerswill JA, Henderson JR, Henderson MN (1996) Distribution of gold and sulphides at Lupin, Northwest Territories – a discussion. Econ Geol 91(5):957–963

Langlais VH, Beucher D, Renard (2008) In the shade of the truncated gaussian simulation, In: Geostats 2008 Ortiz JM, Emery X (eds) Proceedings of the 8th International Geostats Conference, held in Santiago Chile from 1_5 December 2008

Lantuéjoul C (1997a) Iterative algorithms for conditional simulations. In: Baafi E et al (eds) Geostatistics Wollongong '96, vol 1. Kluwer, Dordrecht, pp 27–40

Lantuéjoul C (1997b) Conditional simulation of object-based models. In: Jeulin D (ed) Advances in theory and applications of random sets. World Scientific, Singapore, pp 271–288

Lantuéjoul C (2002a) Geostatistical simulation: models and algorithms. Spinger, Berlin, 250 pp

Lantuéjoul C (2002b) Conditional simulation of spatial stochastic models. In: Talbot H, Beare R (eds) Proceedings of ISMM2002. pp 327–335

Le Loc'h G (1989) An efficient strategy for combining the permeabilities: practical application on a simulated reservoir. In: Armstrong M (ed) Geostatistics, vol 2. Kluwer, Dordrecht, pp 557–568

Le Loc'h G, Beucher H, Galli A, Doligez B (1994) Improvement in the truncated gaussian method: combining several Gaussian functions. In: Proceedings of ECMOR IV, 4th European conference on the mathematics of oil recovery, Roros, Norway, 13p

Le Loc'h G, Galli A (1997) Truncated plurigaussian method: theoretical points of view. In: Baafi E et al (eds) Geostatistics Wollongong '96, vol 1. Kluwer, Dordrecht, pp 211–222

Le Maux T, Fanta O, Sarda S, Godail L (2005) What is the impact of fracture modelling on reservoir performances & reservoir simulation: examples. SPE 97383, presented at the SPE Latin American & Caribbean petroleum engineering conference, held in Rio de janeiro, Brazil, 20–23 June 2005

Le Ravalec-Dupin M, Roggero F, Froidevaux R (2004) Conditioning truncated gaussians realizations to static & dynamic data, SPE 84944. SPE J 9(4):475–480

Lemouzy PM, Eschard R, Beucher H (1991) An integrated approach for evaluation of production scenarios. SPE 22906, 66th Annual technical conference and exhibition of the SPE, pp 209–219

Liu N, Oliver DS (2003a) Conditional simulation of truncated random fields using gradient methods. Proceedings of IAMG annual meeting Portsmouth, UK
Liu N, Oliver DS (2003b) Automatic history matching of geologic facies. SPE 84594, presented at the SPE annual technical conference & exhibition, held in Denver, CO, 5–8 Oct 2003
Liu N, Oliver DS (2005a) Critical evaluation of the Ensemble Kalman Filter on history matching of geologic facies. SPE 92867 presented at the SPE 2005 Reservoir simulation symposium held in Houston Texas, 31 Jan–2 Feb 2005
Liu N, Oliver DS (2005b) Ensemble Kalman filter for automatic history matching of geologic facies. J Petrol Sci Eng 47:147–161
Lorentzen RJ, Naevdal G, Vallès B, Berg AM, Grimstad AA (2005) Analysis of the Ensemble Kalman filter for estimation of permeability and porosity in reservoir models, SPE 96375, presented at the SPE Annual Technical Conference and Exhibition, Dallas, 9–12 October, doi: 10.2118/96375-MS
Mariethoz G, Renard P, Jaquet O, Cornaton F (2006) High resolution stochastic modelling of aquifer, examples for a contaminant migration problem. 4th Swiss Geoscience Meeting, Bern
Mariethoz G, Renard P, Cornaton F, Jaquet O (2009) Truncated plurigaussian simulations of aquifer heterogeneity. Ground Water 47(1):13–24
Matheron G (1968) Schéma booléen séquentiel de partition aléatoire. Internal report N-83, CMM, Fontainebleau
Matheron G (1975) Random sets and integral geometry. Wiley, New York
Matheron G (1987) Suffit-il, pour une covariance, d'etre de type positif? Sci Terre Inf Geologique 26:51–66
Matheron G (1989) The internal consistency of models in geostatistics. In: Armstrong M (ed) Geostatistics, vol 1. Kluwer, Dordrecht, pp 21–38
Matheron G (1993) Une conjecture sur la covariance d'un ensemble aléatoire. Cahiers de Geostatistique 3:107–113
Matheron G, Beucher H, de Fouquet C, Galli A, Guérillot D, Ravenne C (1987) Conditional simulation of the geometry of fluvio-deltaic reservoirs. SPE 1987 Annual technical conference and exhibition, Dallas, Texas, pp 591–599. SPE 16753
Matheron G, Beucher H, de Fouquet C, Galli A, Ravenne C (1988) Simulation conditionnelle à trois facies dans une falaise de la formation du Brent. Sci Terre Inf 28:213–249
Mathieu Y, Verdier F, Houel P, Delmas J, Beucher H (1993a) Reservoir heterogeneity in fluvatile Keuper facies: a subsurface and outcrop study. In: Eschard R, Doligez B (eds) Subsurface reservoir characterization from outcrop observations. Editions Technip, Paris, pp 145–160
Mathieu Y, Verdier F, Beucher H (1993b) Reservoir heterogeneity in fluviatile Keuper facies: a subsurface and outcrop study. In: Eschard R et al (eds) Subsurface reservoir characterization from outcrop observations. Technip, Paris, pp 145–160
Metropolis N, Rosenbluth AW, Teller AH et al (1953) Equations of state calculations by fast computing machines. J Chem Phys 21:1087–1091
Meyn SP, Tweedie RL (1993) Markov chains and stochastic stability. Springer, Berlin, 550 pp
Molchanov I (1997) Statistics of the Boolean model for practitioners and mathematicians. Wiley, Chichester, 160 pp
Morelon IF, Doligez B, Guérillot D, Rahon D, Touffait Y (1991) An application of a 3D geostatistical imaging to reservoir fluid flow simulations. In: Proceedings of the sixth SPE petroleum computer conference. pp 223–231, SPE 22312
Moulière D (1998) Intégration d'information sismiques pour la simulation de réservoirs. Thèse de Docteur de l'Ecole des Mines de Paris, pp 182
Moulière D, Beucher H, Hu L-Y, Fournier F, Terdich P, Melchiori F, Griffi G (1997) Integration of seismic derived information in reservoir stochastic modelling using the truncated gaussian approach. In: Baafi E et al (eds) Geostatistics Wollongong '96, vol 1. Kluwer, Dordrecht, pp 374–385
Mubarak SM, Ali AA, Pham TR, LeMaux T, Colomar FM (2009) Validation of fracture lineaments with dynamic well data, improves history matching of a dual porosity-permeability

model. SPE 119643, presented at the 2009 SPE Middle East oil & gas show & conference, held in Bahrain, 15–18 Mar 2009

Naevdal G, Johnsen LM, Aanonsen SI, Vefring EH (2005) Reservoir monitoring and continuous model updating using Emsemble Kalman Filter. SPE J 10(1):66–74, SPE-84372-PA, doi: 10.2118/84372-PA

Normando NM, Remacre AZ, Sancevero SS (2005) The study of plurigaussian simulations' lithotype rule in reservoir characterization process. SPE 94949, presented at the SPE Latin American & Caribbean petroleum engineering conference, held in Rio de Janeiro, Brazil, 20–23 June 2005

Numelin E (1984) General irreductible Markov chains and no negative operators. Cambrdige University Press, Cambrdige

Oliviera RM, da S. Ribeira NM Jr, Schroeder PR, Johann LFC Jr, Steagal DE, Kerber PA, Carvalho MRJ (2005) Using seismic attributes to estimate net thichness in pinch-out areas – Marlim deep water turbidite oilfield, Campos Basin. SPE 94913, presented at 2005 SPE Latin American & Caribbean petroleum engineering conference held in Rio de Janeiro, Brazil 20–23 June 2005

Pelgrain de Lestang A, Cosentino L, Cabrera J, Jimenez T, Bellorin O (2002) Geologically oriented geostatistics: an integrated tool for reservoir studies. SPE 74371, presented at the SPE international petroleum conference & exhibition, held in Villahermosa, Mexici, 10–12 Feb, 2002

Pontiggia M, Ortenzi A, Ruvo L (2010) New integrated approach for diagensis characterization and simulation. SPE 127236, presented at the SPE North African technical conference & exhibition held in Cairo, Egypt, 14–17 Feb 2010

Ravenne C, Beucher H (1988) Recent developments in description of sedimentary bodies in a fluvio-deltaic reservoir and their 3D conditional simulations. In: Proceedings of the 63rd annual technical conference and exhibition of SPE. pp 463–476. SPE 18310

Ravenne C, Eschard R, Galli A, Mathieu Y, Montadert L, Rudkiewicz JL (1987) Heterogeneities and geometry of sedimentary bodies in a fluvio-deltaic reservoir. SPE 1987 Annual technical conference and exhibition, Dallas, Texas. Vol X. pp 115–122. SPE 16752

Ravenne C, Galli A, Beucher H, Eschard R, Guérillot D (1991) Outcrop studies and geostatistical modelling of a middle Jurassic Brent Analogue. In: Imarisio G et al (eds) The European oil and gas conference, A multidisciplinary aproach in exploration and production R&D. Chapman & Trotman, London, pp 497–520

Ravenne C, Galli A, Preat A (1994) Reservoir modeling: the reference model: Poulseur/CBA. In: Helbig K (ed) Modeling the earth for oil exploration. Pergamon, Brussels, pp 91–96

Ravenne C, Galli A, Doligez B, Beucher H, Eschard R (2002) Quantification of facies relationships via proportion curves. In: Armstrong M et al (eds) Geostatistics Rio 2000. Kluwer, Dordrechts, Holland, pp 19–40

Remacre A, Simon V (1998) Conditional plurigaussian simulation: a case-study. In: Cahiers de Géostatistique, Ecole des Mines de Paris, vol 6. 17–28

Renard D, Beucher H, Doligez B (2008) Bi-plurigaussian simulation model with heterotopic information. In: Ortiz JM, Emery X (eds) Geostats 2008. Proceedings of the 8th international geostats conference, held in Santiago, Chile, 1–5 Dec 2008

Riquelme R, Le Loc'h G, Carrasco P (2008) Truncated gaussian & plurigaussian simulations of lithological units in Mansa Mina Deposit. In: Ortiz JM, Emery X (eds) Geostats 2008. Proceedings of the 8th international geostats conference, held in Santiago, Chile, 1–5 Dec 2008

Robert CP (1995) Simulation of truncated normal variables. Stat Comput 5:121–125

Robert CP (1996) Méthodes de Monte Carlo par Chaines de Markov. Economica, Paris, 340 pp

Roberts GO, Polson NG (1994) On the geometric convergence of the Gibbs sampler. J R Statist Soc B 56(2):377–384

Roberts GO, Sahu SK (1997) Updating schemes, correlation structure, blocking and parametrization for the Gibbs sampler. J R Statist Soc B 59(2):291–317

Rockafellar RT (1970) Convex analysis, vol 28, Princeton mathematical series. Princeton University Press, Princeton, NJ, p 29

Rondon O (2009) A look at plurigaussian simulation for a nickel laterite deposit. Presented at the 7th international mining & geology conference, held in Perth, WA 17–19 Aug

Roth C (2000) Incorporating information on edge effects when simulating lithofacies. Math Geol 32(3):277–300

Roth C, Armstrong M, Galli A, Loc'h G (1998) Using plurigaussian simulations to reproduce lithofacies with contrasting anisotropies. In: APCOM '98, London, IMM, pp 201–213

Rudkiewicz JL, Guérillot D, Galli A (1990) An integrated software for stochastic modeling of reservoir lithology and properties with an example from the Yorkshire middle Jurassic, In: North Sea Oil and Gas Reservoirs II. The Norwegian Institute of Technology, Graham and Trotman, pp 399–406

Roberts GO, Sahu SK (1997) Updating schemes, correlation structure, blocking and parametrization for the Gibbs sampler. J R Statist Soc B 59(2):291–317

Seigneurin A, Muller D, Galli A, Ravenne C (1993) Optimization of the well-spacing with a geostatistical model Tunu field – Mahakam area. In: Proceedings of Indonesian Petroleum Association, 22nd Annual Convention, 17 p

Skjervheim J-A, Ruud BO, Aanonsen SI, Johansen TA (2007) Incorporating 4D seismic data in reservoir simulation model using Ensemble Kalman Filter. SPE J 12(3):282–292, SPE-95789-PA, doi: 10.2118/95789-PA

Skvortsova T, Armstrong M, Beucher H, Forkes J, Thwaites A, Turner R (2000) Applying plurigaussian simulations to a granite-hosted orebody. In: Kleingeld W, Krige DG (eds) Geostats 2000 Cape Town. Proceedings of the 6th international Geostatistics Congress, held in Cape Town 10–14 April 2000, pp 904–911

Skvortsova T, Armstrong M, Beucher H, Forkes J, Thwaites A, Turner R (2002) Simulating the geometry of a granite-hosted uranium orebody. In: Armstrong M et al (eds) Geostatistics Rio 2000. Kluwer, Dordrechts, Holland, pp 85–100

Thomas P, Le Ravale-Dupin M, Roggero F (2006) History matching reservoir models simulated from the plurigaussian method. SPE 94168, presented at the SPE Europe/ EAGE annual conference held in Madrid, Spain 13–16 June 2005

Tierney L (1994) Markov chains for exploring posterior distributions, with discussion. Ann Stat 22:1701–1762

Tierney L (1996) Introduction to general state-space Markov chain theory. In: Gilks WR, Richardson S, Spiegelhalter DJ (eds) Markov Chain Monte Carlo in practice. Chapman & Hall, Boca Raton

Touffait Y, Beucher H, Guérillot D (1990) A 3D integrated structure for computer aided reservoir characterization. In: Proceedings of the 5th SPE petroleum computer conference. pp 171–180. SPE 20348

Van Buchem FS, Doligez B, Eschard R, Lerat O, Grammer M, Ravenne C (2000) Stratigraphic Architecture and Stochastic Simulation of a Mixed silico-clastic/carbonated platform (Pennsylvanian, paradox basin, USA. In: Homewood P, Eberli GP (eds) Genetic stratigraphy, vol 28, Bulletin de Recherche Exploration-Production memoir., pp 109–128

Volpi B, Galli A, Ravenne C (1997) Vertical proportion curves: a qualitative and quantitative tool for reservoir characterization. In: Memorias del I Congreso Latinoamericano de Sedimentologia. Soc Venezolana de Geol. Tomo 2, pp 351–358, Nov 1997

Xu C, Dowd PA (2008) Plurigaussian simulation of rock fractures. In: Ortiz JM, Emery X (eds) Proceedings of the Eighth international geostatistics congress. vol 1, pp 41–50 pub. Gecamin, Santiago, ISBN 978-956-8504-17-5

Xu C, Dowd PA, Mardi K, Fowler R (2006) A flexible true plurigaussian code for spatial facies simulations. Comput Geosci 32:1629–1645

Yarus JM, Chambers RL (2006) Practical geostatistics – an armchair view for petroleum reservoir engineers, J Petroleum Technology Nov 2006, p 78–87

Yosida K (1978) Functional analysis, 5th edn. Springer, Berlin

Zhao Y, Reynolds AC, Li G (2008) Generating facies maps by assimilating production data and seismic data with the Ensemble Kalman filter. SPE 113990 presented at 2008 SPE/DOE improved oil recovery symposium held in Tulsa, Oklahoma, 19–23 April, 2008

Index

A
Aggrading, 123
Algal mounds, 1, 9, 127–129, 132, 133
Algorithm, acceptance/rejection, 111–112
Allocyclic, 121
Alluvial gravels, 86
Analogue, 122
Anastomosed channels, 61
Andesitic flows, 65
Anisotropy, 100, 101, 126, 130, 132, 150, 155–158, 160
Aperiodic, 33
Aquifer
- contamination, 8
- heterogeneous, 8

B
Back barrier, 61, 130
Back reef, 129, 130
Banded iron formation (BIF), 62
Basin-ward, 130
Bayes theorem, 32
BIF*See* Banded iron formation (BIF)
Bioherms, 82, 126, 127
Blocking factor, 28–30
Bochner's theorem, 49
Boolean, 121, 126
Brazil, 134
Breccia, 65, 145, 146
Burn-in period, 116–119

C
Cajigar 2 succession, 65
Capillary pressure, 124
Carbonates, 11, 137
Categorical variables, 12, 16
Cemented horizon, 130
Chalco-bornite, 65
Chalco-pyrite, 65
Channel, 6, 61, 63, 64, 121, 123
Chile, 8, 10, 12, 65, 86, 107, 121, 138, 145
Chronostratigraphic marker, 50, 65, 67, 68
Chuquicamata, 144
Clastic fluvial, 134
Cleveland, 61
Coastal plain, 126
Cokriging, 23–25, 65
Compact, 28, 30, 33, 34, 37, 38
Conditional expectation, 11, 31, 32, 113
Conditional simulation, 6, 15, 139, 142
Connectivity, 45, 72, 133, 134
Constraints
- categorical, 23–24
- hard, 3
- interval, 27, 111
- seismic, 24–26
- soft, 3

Contacts, forbidden, 81
Convergence, 9, 27–38, 112, 116, 117
Convex, 37–38
Convolution, 12, 22, 24, 25
Copper, 8, 10, 65, 66, 85, 86, 138, 144–147
Correlation, 6, 23, 25, 26, 65, 79, 81–83, 90–92, 105, 108, 110, 112, 116–118, 121, 122, 130, 135, 136, 138, 154, 160
Covariance
- centred, 42, 54, 87
- 2 facies, 23, 44, 47–48
- 3 facies, 22, 45–47
- noncentred, 16, 20, 42, 52, 54, 55, 88, 104, 105
- symmetrical, 105

Crevasse splay, 21, 23, 123
Cross variogram, 11, 12, 24, 25, 42–45, 65, 87–90, 93–96, 156, 157

D
Decomposition
  Cholesky, 28, 112
  LU, 107, 111, 112
Deltaic reservoir, 61, 64, 122
Depositional cycles, 68
Derivative, 6, 12, 21, 22, 92, 97
Diagenesis, 8, 129–131, 134–136, 138
Diamond pipe, 8
Distance metric, 33–35

E
Eigenvalues, 34, 35, 37
Ellerbeck, 61, 62, 68
Ensemble Kalman filter (EnKF), 8
Environmental science, 8
Ergodicity, 30
Erosion, 57, 67, 123
Estuarine channel, 123
External drift, 72

F
Flanking beds, 127, 129
Flooding surface, 68
Flood plain, 62, 64, 68, 123
Fluid flow
  multiphase, 9
  simulator, 3, 8, 125
Fluvial channels, 123
Fluvial reservoir, 6, 64, 121, 123
Fluvial sand sheet, 122
Foreshore, 62, 68, 121, 126
Fracture
  modelling, 7
  network, 132
Freeware, 6, 10, 149–161

G
Galerkin approximation, 35
Gamma process, 18–20
Gas, 125
Gauss-seidel, 28
Geological boundaries, 7
Gibbs sampler, 7, 9, 27–30, 35–38, 107–120
Gold, 62–63
Gradual deformation, 8, 9

H
Hammersley clifford, 35–36
Hastings metropolis, 112
Heterotopic, 10, 134–138
History matching, 3, 8–9, vi
Hydrodynamic, 138–139
Hydrology, 7

I
Index, 13, 45, 113, 134–136
Installation
  linux, 149–150
  windows, 149–150
Interfingering, 121, 122, 126
Inter-mound, 127
Irreducible, 32–33

J
Jurassic age, 61

K
Kalman filter, 8
Karstification, 127
Kernel operator, 31
Khazakstan, 138
Kriging, indicators, 11

L
Lagoon, 62, 68, 129, 130
Leaching, 138
Lithotype rule, 83, 126–131, 139, 142
Lupin orebody, 63

M
Mantos blancos mine, 65
Markov chain, 30–38, 112, 116, 118
Mass balance, 9
MCMC, 27, 112, 116
Meandering channels, 21, 22, 61
Mining, 1, 7, 8, 10, 11, 59, 65, 79, 112, 138–139
Minorisation condition, 37, 38
Miocene, 129
Monte carlo, 112
Mudstone, 62, 63, 68, 123
Multimodal distribution, 21
Multi-point geostatistics, 48

N
Nested simulations, 121, 123, 135
Nickel laterite deposit, 8
Non sedimentary, 138, 144–147
Non stationarity, 9, 13, 59, 63–67, 72, 77, 84, 94, 97, 101, 103, 122, 154
North sea reservoirs, 9

O
Object-based, 121–123
Offshore, 7, 121, 126
Oil, 1, 3, 7–9, 21, 23, 59, 67, 125, 134, 138
Onlap, 72, 125
Outcrop study, 123
Oxides, 85, 86
Oxidizing, 138, 139, 141, 142

P
Paradox basin, 9, 127–129
Partition, 7, 11, 17, 27, 34, 37, 77–83, 154
Pennsylvanian age, 1, 127
Periodic, 33
Permeability, 1, 2, 8, 20, 122, 124, 125, 130
Petrophysical data, 2, 3, 21
Pixel based, 8
PluriDemoSet, 10, 149, 158–161
PluriDemoSimu, 10, 149–151, 153, 155, 156,158
PluriDemoVario, 10, 149, 154, 155, 157
Porosity, 1, 2, 8, 9, 57, 124, 125, 129, 130
Porous medium, 57
Porphyry copper, 8, 10, 85, 86, 138, 144–147
Practical range, 97–100, 113, 116, 117
Prograding, 1, 9, 126, 127
Proportion curve
  horizontal, 65, 66, 72, 122
  vertical, 9, 59–69, 72, 77, 122, 126, 127, 130, 131, 139, 143, 153
Proportion matrix, 64, 72, 135, 139, 140, 146

Q
Quartz veins, 62

R
Radomiro tomic, 86
Random function (RF)
  correlated, 6
  indicator, 11, 41
  multivariate, 11
  shifted, 6
  stationary, 44
  translated, 6
Random set, 11, 41, 44, 45, 48, 49, 57, 149, 158, 160
Ravenscar, 60–62, 68
Recovery process, 2, 3, 125
Reducing, 138
Reef, 9, 129–134
Reference level, 50, 59, 62, 67–72, 124–125, 145
Regressive-transgressive, 122
Reservoir characterisation, 2, 122
Reservoir type
  carbonates, 1, 127
  deltaic, 61, 64
  fractured, 7, 8
  north sea, 9
  reef, 9, 129
  siliclastic, 8, 123
  turbidites, 123
Rock type rule, higher dimensional, 85–86
Roll front uranium deposit, 8, 10, 26, 138

S
Saltwick, 61, 62
Sample locations, 7, 12, 13, 15, 23, 25
Sandstone, 1, 6, 59, 60, 62, 63, 65, 68, 72, 77, 84, 123, 127, 129, 138
San Juan, 1, 2, 127
Scarborough formation, 60, 61
Sea level changes, 61
Seismic data, 2, 16, 65, 72, 124
Seismic inversion, 8
Sequence stratigraphy, 3, 13, 121
Sequential gaussian, 26, 107
Shale, 1, 6, 59, 60, 84, 85, 123, 139–141
Shift
  deterministic, 26
  stochastic, 26
Shoreface, 62, 121, 122, 126
Siliclastic reservoir, 8, 123
Simple kriging (SK), 18, 22, 25, 49, 108, 113
Sinuosity, 123
Spectral radius, 28–30
Stationarity
  horizontal, 51
  vertical, 50, 101
Stockwork, 145, 146
Subjective, 72
Subsidence, 67
Sulphide, 1, 62, 66, 85, 86

T
Target points, 11, 13, 18, 39
Tectonic deformation, 132
Thickness, 65, 71, 72, 123
Thin section, 134
Thresholds, 4–7, 9, 18, 62–63, 73–86, 89, 93, 95, 118, 132, 151, 152, 154, 160
Toplap, 72, 125
Total variation distance, 30, 37
Transformation
  linear, 160
  non linear, 18
Transgressive, 62, 68, 122
Transition probability
  2nd type, 51–53
  1st type, 49–51
Triangle inequality, 48, 56
Turning bands, 107

U
Unconformity, 67, 68
Updating strategy, 115–116
Upscaling, 3, 125
Uranium deposit, 8, 10, 26, 138

V
Variograms
  cardinal sine, 150, 152, 155
  experimental, 93–95
  exponential, 98
  2 facies, 44, 84, 112, 149, 158
  3 facies, 152, 154, 158
  fitting, 96, 101–104
  gaussian, 95–96
  horizontal, 94, 122
  indicator, 95–96, 99
  levels, 94
  nugget effect, 74, 112, 156
  power function, 48
  range, 122, 124, 133
  spherical, 49
  vertical, 97, 141
Vertical proportion curves, 9, 59–65, 67, 72, 77, 122, 131, 139, 143

W
Walther's law, 121
Washouts, 23
Washover fan, 61
Waste, 63, 65, 145–147
Well logs, 3, 124

Y
Yorkshire, 60, 63

GPSR Compliance

The European Union's (EU) General Product Safety Regulation (GPSR) is a set of rules that requires consumer products to be safe and our obligations to ensure this.

If you have any concerns about our products, you can contact us on ProductSafety@springernature.com

In case Publisher is established outside the EU, the EU authorized representative is:

Springer Nature Customer Service Center GmbH
Europaplatz 3
69115 Heidelberg, Germany

Batch number: 10614709

Printed by Printforce, the Netherlands